ELEMENTARY SCIENCE

OF METALS

DATE DUE			

THE WYKEHAM SCIENCE SERIES

for schools and universities

General Editors:

Professor Sir Nevill Mott, F.R.S.
Cavendish Professor of Physics
University of Cambridge

G. R. Noakes
Formerly Senior Physics Master
Uppingham School

To broaden the outlook of the senior grammar school pupil and to introduce the undergraduate to the present state of science as a university study is the aim of the Wykeham Science Series. Each book seeks to reinforce this link between school and university levels, and the main author, a university teacher distinguished in the field, is assisted by an experienced sixth-form schoolmaster.

ELEMENTARY SCIENCE
OF METALS

J. W. Martin – University of Oxford

DISTRIBUTED BY
SPRINGER-VERLAG NEW YORK INC.
175 Fifth Avenue New York, NY 10010
ISBN #91045-X

WYKEHAM PUBLICATIONS (LONDON) LTD
(A subsidiary of Taylor & Francis Ltd)
LONDON & WINCHESTER
1969

PREFACE

By studying metallurgy, we bring together several scientific and technological disciplines, and this synthesis springs from the great scientific interest of metals and also from their immense practical value to mankind in our civilization. After an Introduction which shows how metallurgy interlinks pure and applied science, the main text confines itself to an exposition of the relationship between the structure (Chapters 1–3) and the properties (Chapters 4–6) of metals. My aim has been to demonstrate to students of the pure sciences (for example to sixth-formers studying physics and chemistry) that a modern approach to the applied science of metallurgy has established it as an intellectually rewarding subject for study not only at school itself (where a number of ' options ' in applied science are becoming available to G.C.E. students) but also for further study after school and also as a possible career.

The level of the text is intended for school sixth-formers, and I would like to express my sincere gratitude to Mr. R. A. Hull for his advice and guidance so patiently given in establishing this level and in clarifying the approach when the book was in preparation: I do, however, take full responsibility for the final product.

Oxford

May 1969   J. W. MARTIN

v

INTRODUCTION

IT has been said that metallurgy is one of the oldest of the arts, but one of the youngest of the sciences. We can define the subject as "the science and art of processing and adapting metals to satisfy human wants", and it is easy to see that metals have played a key role in the development of civilization. Stages in the history of mankind have been named the Bronze Age and the Iron Age, for example, and in the Nuclear and Space Age of today the advance of our technology depends critically upon the contribution of the metallurgist.

There are over 20 000 metals and alloys in present use, and their number is constantly increasing, due to the demands of our society, and it is the great metal-producing and metal-using industries which primarily determine the wide scope of activities of the metallurgist. Professor F. D. Richardson of the Imperial College of Science and Technology, London, has graphically analysed the 'anatomy of metallurgy' today by starting with the industrial applications of metallurgy and working back to basic principles to demonstrate the truly interdisciplinary nature of the subject. Professor Richardson expressed the structure of metallurgy diagrammatically in the two triangles shown in figs. P.1 *a* and *b*.

Applied metallurgy (fig. P.1 *a*) can be divided into three main sections. There is *extractive metallurgy* which extends from ore pre-treatment through extraction and refining of the metals to solidification. There is *mechanical metallurgy*, which consists of casting, shaping and metal treatments of various types. There is *design of materials*, which is concerned with the behaviour of industrial alloys and other materials of metallurgical significance such as ceramics and refractory materials.

These three applied sections of the subject would be devoid of any theoretical foundation without the three *basic* sections represented in fig. P.1 *b*. *Chemical metallurgy* is concerned with the obtaining of a metal from its ore, and also with the reverse of this process, namely, corrosion, when a metal returns to a state of chemical combination. *Physical metallurgy* seeks an understanding of all the physical and mechanical properties of metals in terms of what is happening at the atomic level. *Process science* is concerned with fluid mechanics and heat and mass transfer in all kinds of dynamic situations relevant to large-scale processing.

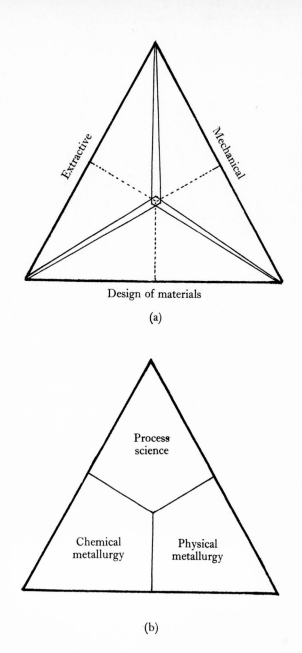

Fig. P.1. The anatomy of metallurgy : (a) The applied sections. (b) The basic sections.

In fig. P.1 a the three applied sections are represented overlying the three basic sections on which they depend: extractive metallurgy based primarily on chemical metallurgy and process science; mechanical metallurgy based primarily upon physical metallurgy and process science; and design of materials based on chemical and physical metallurgy. In addition the applied sections have been drawn with an overlap to stress that none of them rests solely on two of the three basic sections.

The choice of a material for a given application depends not only upon the effectiveness of that material in the application but also upon its cost. The pattern of usage of many metals has changed over the years as cheaper methods of production have been developed. For example, when aluminium was first produced in France in the mid-nineteenth century by the reduction of a compound by means of sodium metal, its cost was comparable with that of the precious metals. It was in fact used for jewellery and for *objets d'art*, but an electrolytic method of reduction was devised in 1889, and now the metal is second only to steel in cheapness and in tonnage demand. In summary, one can say therefore that metallurgy is based upon the *pure sciences* of chemistry, physics and mathematics, and upon the *applied sciences* of engineering and of economics.

Although it can be seen that metallurgy forms a very powerful combination of disciplines to study, it is obvious that an introduction can only be given within the scope of the present text to a very limited aspect of the subject as a whole. The various topics discussed in the following chapters form part of the basic principles of *physical metallurgy*. The emphasis has been placed upon the structures of solid metals and alloys and how these structures are related to their properties, illustrating the principles as far as possible with examples of practical significance. This does not constitute an introduction to the whole field of physical metallurgy—for example, solidification processes are dealt with only briefly, and no consideration has been given to electrical or magnetic properties—but it is hoped that a reader versed in school physics and chemistry will gain an insight into one corner of the subject, to use a term appropriate to the Richardson triangle!

Supplementary reading list
Metals in the Service of Man : A. Street and W. Alexander (Pelican).
Man, Metals and Modern Magic : J. G. Parr (American Society for Metals).

CONTENTS

the structure of metals

1.1. *Introduction*

WHAT do we mean by the *structure* of a metal? At its most basic, the term can apply to the constitution of the individual atoms of the material : for example, their electronic structure can be established and correlated with their position in the Periodic Table, which in turn is linked with the *chemical* properties of the element. Again, the bonding together of aggregates of atoms in a metal can be considered, and the *physical* properties of metals can be interpreted in terms of the ' metallic bond ' the nature of which we shall discuss later. The physical metallurgist, however, is often concerned with yet another aspect of the structure of metals, which is that sometimes observed with the naked eye—but more usually under a microscope—and is called the ' microstructure '. One of the basic aims of the physical metallurgist is to try to correlate the microstructure he sees with the *mechanical* properties he measures, so that, by understanding the relationship between these two factors, he will know how to prepare materials with the properties required by design engineers.

Many mineral substances are obviously *crystalline*, because their external surface is covered with the faces of crystal planes. One definition of a crystal (and it is not a very satisfactory one, as we shall see) is " a homogeneous solid bounded by naturally formed plane faces ", and early crystallographers based their studies on the external appearance of crystals. They deduced that the regular arrangement of the external faces of these solids was related to the way in which the matter of the crystal was assembled as it grew. Crystallinity, therefore, implies that the constituent atoms, ions or molecules of a solid are arranged in a regular geometric pattern, and this ' internal structure ' will possess a symmetry similar to the external symmetry of the crystal.

Examination of the surface of any familiar metallic object—a coin, watch or key for example—gives no clue as to whether or not the metal is crystalline. The shape of its surface is determined by the profile of a mould (if it has been cast from the liquid state) or by the forces exerted in a forging process (if it has been mechanically shaped), so it is the regularity of its *internal structure* (i.e. the arrangements of its atoms) which we must consider if we are to define if such an object is crystalline or not. In fact all metals *are* crystalline, and their atoms are arranged in specific geometric patterns ; the most

1

striking early evidence for this fact was produced in 1808, when Alois de Widmannstätten examined a meteorite, which consisted of an alloy of iron and nickel. He polished the surface of a section through the meteorite, and then chemically attacked it in a dilute acid, and thereupon discovered a beautiful geometric pattern (fig. 1.1) of

Fig. 1.1. Section of an iron–nickel meteorite prepared by Alois de Widmannstätten, published in 1820. The original is a direct typographical imprint from the etched surface. (Reproduced approximately two-thirds of original size.)

regularly arranged crystals, easily visible to the naked eye. It was therefore appreciated by mineralogists in the nineteenth century that, in common with other mineral substances, metals are crystalline in character, although more direct experimental evidence to verify the presence of an internally symmetric structure in crystals did not emerge until the beginning of the present century, after the discovery of X-rays.

The arrangement of the atoms within a material is primarily determined by the strength and the directionality of the *interatomic bonds*. Only the inert gases (He, Ne, Ar, Kr, Xe and Rn which have closed outer-electron shells) exist as individual atoms: elements usually exist as molecules (comprised of two or more atoms bonded together) or as crystals (where many atoms or molecules take up an ordered array), and to understand any of the properties of matter we must first survey the bonding behaviour of atoms in their formation of such molecules or crystals.

1.2. *Bonding in solids*

We can understand qualitatively why an atomic bond is strong or weak, and directional or non-directional, from a knowledge of the energies and distribution of the bonding electrons with respect to the positively charged ions in the solid. Three types of strong (or primary) bonding can be distinguished, namely ionic, covalent and metallic bonding, and they differ from each other in the way in which the bonding electrons are localized in space. In fact these three types of bond are limiting cases, and a whole range of intermediate bonding situations also exist in solids. Besides these primary bonds, relatively weak secondary attractions exist in solids and are termed van der Waals forces.

The nature of the bonding in non-metals is usually discussed at length in most elementary textbooks on chemistry, so that we will only consider these briefly here, in order that their properties may be compared to those of the metallic bond (§ 1.4).

1.2.1. *Ionic bonding*

Ionic bonding is primarily an electrostatic attraction between positive and negative ions which are formed from the free atoms by the loss or gain of electrons. Elements such as metals can easily dissociate into positive ions and free electrons and are classified as *electropositive* elements, whereas the atoms of the elements such as oxygen, sulphur or the halogens tend to acquire electrons and become negative ions and are known as *electronegative* elements. If free atoms of an electropositive element and an electronegative element are brought together, they are thus chemically attracted and positive and negative ions will be formed, each consisting of a positive nucleus surrounded by a cloud of electrons. These ions will be pulled together by electrostatic attraction until the electron clouds of the two ions start to overlap, which gives rise to a repulsive force. The ions thus adopt an equilibrium spacing at a distance apart where the attractive and repulsive forces just balance each other.

In an ionically bonded solid, each ion with charge of one sign attracts *all* neighbouring ions with charge of the opposite sign, so that the ionic bond is *non-directional*. Each ion tends to be surrounded by as many ions of opposite charge as can touch it simultaneously—the actual number of surrounding ions being dictated firstly by the relative sizes of the two types of ion and secondly by the necessity of

3

maintaining the solid in an electrically neutral state. Figure 1.2 shows a diagram of the structure of a sodium chloride crystal : the radius of the sodium ion is so much smaller than that of the chlorine ion that there is room round the latter for twelve or more sodium neighbours. Electrical neutrality, however, must be preserved by

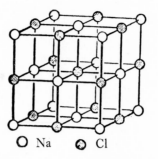

O Na   ◉ Cl

Fig. 1.2. The crystal structure of sodium chloride; the lines do *not* represent chemical bonds, but are drawn to emphasize the cubic array and to illustrate that each Na^+ ion is surrounded by six Cl^- ions and that each Cl^- ion is surrounded by six Na^+ ions.

maintaining equal numbers of sodium and chlorine ions, and since only six chlorine neighbours can be accommodated round each sodium ion, each chlorine ion has in turn only six sodium neighbours.

It is obvious that the molecule NaCl has no existence in the structure, and that the number of neighbours possessed by a given ion is a geometrical property of the structure. Many of the physical properties of ionic crystals may be accounted for qualitatively in terms of the characteristics of the ionic bond : for example, ionic crystals are non-conductors of electricity since the electrons are all firmly bound to individual ions.

1.2.2. *Covalent bonding*

Ionic bonds are possible as long as the two elements concerned are from opposite sides of the Periodic Table, so that one tends to be electropositive and the other electronegative, as in the case of sodium (Group 1) and chlorine (Group 7) in the formation of sodium chloride. However, when the elements concerned are adjacent to one another in the Periodic Table, or when a crystal or molecule of a single element is formed, bonds of the covalent or of the metallic type must be involved.

The simplest examples of stable covalent bonds are those formed between non-metallic atoms like carbon, nitrogen, oxygen, fluorine and chlorine. For example, the two chlorine atoms in the molecule of chlorine, Cl_2, each with an outer shell of seven electrons can both attain the stable configuration corresponding to a filled outermost

4

shell of eight electrons by a sharing of two electrons in a way which may be written :

$$:\overset{..}{\underset{..}{Cl}}\cdot + \cdot\overset{..}{\underset{..}{Cl}}: \rightarrow :\overset{..}{\underset{..}{Cl}} : \overset{..}{\underset{..}{Cl}}:$$

where the dots represent the electrons in the outermost shell. The two chlorine atoms are thus joined together so that their electron clouds interlock, and a Cl_2 molecule is formed in which each nucleus has a complete shell at various instants of time. In general, more than one electron can be shared in this way, which is the characteristic feature of covalent bonding, in contrast to the transfer of electrons in ionic bonding, and obviously each atom should have one electron shell at least half-filled for a strong covalent bond to be able to form.

A further way in which the covalent bond differs from the ionic bond is in the fact that the several bonds from a polyvalent atom usually have a characteristic arrangement in space. For example, the four covalent bonds from a carbon atom are always found to be directed towards the corners of a regular tetrahedron (fig. 1.3 a), and the

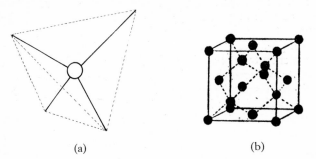

(a) (b)

Fig. 1.3. (a) Showing the tetrahedral bond directions for a carbon atom. (b) Showing the crystal structure of diamond, drawn to illustrate the cubic array.

structure of diamond (fig. 1.3b) provides an ideal example of a covalently bonded crystal. Each carbon atom is surrounded by only four others and the neighbours are all arranged symmetrically at the corners of a regular tetrahedron. Two points about the diamond crystal call for special mention : firstly, the structure is a very 'open' one. If space is to be filled by packing together similar spheres, geometrically *twelve* is the maximum number of spheres that can be packed around a central sphere so that they all touch it, and this arrangement is typical of the crystal structure of many metals where there is a tendency for each atom to surround itself by the largest number of neighbours geometrically possible (see figs. 1.7 and 1.8). In diamond, although all the atoms are equivalent, there is no such tendency, and each atom is surrounded only by that number of atoms

2

5

to which it can be linked by its four covalent bonds, the four shared electrons giving a closed shell. The second point about the diamond crystal is that each atom is individually 'chemically' bonded to its neighbours in a similar way to that in which the two chlorine atoms in a molecule of chlorine are bonded $: \overset{..}{\underset{..}{Cl}} : \overset{..}{\underset{..}{Cl}} :$, so a diamond crystal can be regarded as a single molecule of indefinite extent.

The strength of the covalent bond may be as great as that of the ionic bond, so that mechanically covalent and ionic crystals have similar properties. The physical properties of covalently bonded compounds do show very wide variations, and melting points and hardnesses range over wide limits. Covalently bonded compounds differ from ionic compounds electrically in that they are non-conductors in the molten state, and this is sometimes regarded as a criterion of the covalent bond.

1.2.3. *Metallic bonding*

The non-metals, which occupy the top right-hand corner of the Periodic Table (which is shown inside the front cover), form about one-sixth of all the elements, and they are characterized by having melting-points and boiling-points below about 500°C, and by their solid and liquid phases not conducting electricity. About two-thirds of all the elements are metals, and a further one-sixth have properties intermediate between those of metals and non-metals.

The simple classical picture of a metal suggests that it consists of an array of positive ions immersed in a 'sea' or 'gas' of electrons, as depicted in fig. 1.4. The attraction between the positive ions and the electron gas gives the structure its coherence, and since each valence electron is not localized between only two ions, as in covalent bonding, metallic bonding is non-directional, and the electrons are more or less free to travel through the solid.

Such electrons are said to be 'delocalized': the diffuse nature of the metallic bonding is responsible for the easy deformability, and the free mobility of the electron 'gas' under the influence of an electric field is responsible for the high conductivity. A very satisfactory qualitative or even quantitative description of some of the physical properties of metals can be obtained from this 'free electron' picture of the structure, but many other properties—particularly those concerned with the motion of electrons within metal crystals—cannot be explained in this simple way. The model has had to be replaced by a picture of electrons as waves occupying definite quantized energy states, but we can still employ the simpler concept in discussing the structures of metals.

Generally the bonding will be more metallic the fewer the valence electrons possessed by an atom, and the more loosely they are held. The bonding will tend to become more covalent in character as the number of valence electrons increases, and as the tightness with which they are held to the nucleus increases, so that they become more localized in space.

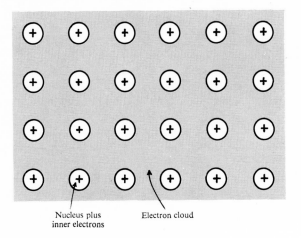

Fig. 1.4. The classical model of a metal crystal: the ions are surrounded by a cloud of negatively charged electrons (approximately one per atom).

1.2.4. *Van der Waals bonding*

This weak secondary bond acts between all atoms or molecules, and it arises from the fluctuations and surges of electronic charge in the atom. At a certain moment there is likely to be found a little bit more electron cloud on one side of a nucleus than on the other side, with the result that the centres of positive and negative charge do not quite coincide. This produces what is called a weak 'dipole', and a force exists between opposite ends of such dipoles in adjacent atoms which tends to draw them together.

It is this van der Waals force which, at low temperatures, permits the inert gases and molecules with electron shells which are effectively full to condense to liquids and to solidify. It is an especially important factor in determining the structure and hence some of the properties of many polymeric materials.

1.3. *The crystal structure of metals*

Having considered the nature of the forces between the atoms in metals, we will now examine the crystalline arrangements to which they give rise. Let us first, however, look briefly at the way in which the use of X-rays has enabled crystallographers to establish the ways in which the atoms are arranged in solids.

7

1.3.1. *X-ray diffraction*

Until about fifty years ago there was no experimental way to find the atomic arrangement in a crystal. Crystallographers had guessed from their studies of the outward form of crystals that the atoms within had a regularly repeated arrangement, but no measurements were possible. In 1912, however, three German physicists, Max von Laue, W. Friedrich and P. Knipping, performed a very important experiment.

Only seventeen years earlier, Roentgen had discovered X-rays, and although physicists were uncertain as to what these rays were, it was suspected that they were a form of light (i.e. electromagnetic radiation) with very short wavelength. Now the phenomena of interference and diffraction were understood in the field of visible light, and they were known to arise for example when light impinges on a grating (fig. 1.5 *a*) provided that (*a*) the grating is regularly periodic of spacing *d*, and (*b*) the wavelength (λ) of the light is of the same order of magnitude as *d*. For a beam incident normally upon a grating, a diffracted beam (fig. 1.5 *a*) at an angle θ will be observed if the path difference (OP)

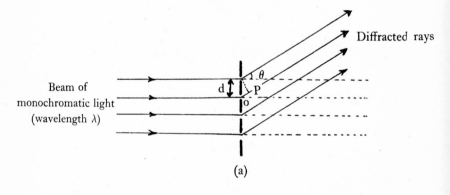

(a)

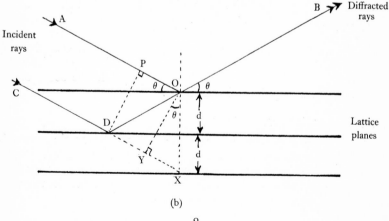

(b)

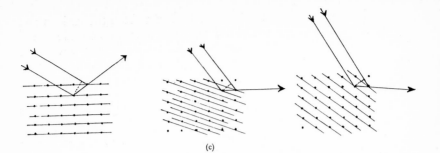

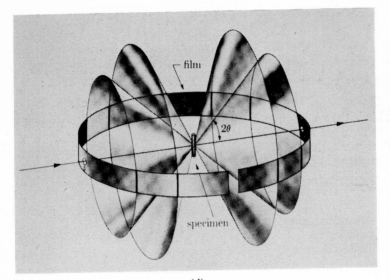

(c)

(d)

Fig. 1.5. (a) Illustrating the diffraction of light by a grating. (b) Illustrating the diffraction of X-rays by the planes of atoms in a crystal. (c) Illustrating the diffraction of X-rays by several possible sets of planes in a crystal. (d) The principle of X-ray powder photographs.

is equal to an integral number of wavelengths (n). This leads to the well-known relationship for diffraction that $n\lambda = d\sin\theta$.

Laue, Friedrich and Knipping showed that a crystal will similarly scatter a beam of X-rays into a large number of separate beams which emerge from the crystal in definite directions : this observation of X-rays being diffracted by a crystal at once confirmed both the fact that X-rays were of wave nature like visible light and also that the atoms in a crystal had a regularly repeated arrangement.

A rigorous discussion of X-ray diffraction in terms of interaction with the three-dimensional array of atoms in a crystal is complex. Bragg has simplified the problem, however, by showing that diffraction is *equivalent* to symmetrical reflection of X-rays from the various sets of planes of atoms in a crystal, provided certain conditions are fulfilled. Figure 1.5 *b* shows a beam of X-rays impinging at an angle θ on a set of crystal planes of spacing *d*. A diffracted beam will appear at an angle θ (i.e. as if reflection had occurred) only if the rays from successive planes reinforce each other. If the wavelength of the X-rays is λ, the path difference for rays diffracted from successive planes (i.e. the extra distance each successive ray has to travel) must be equal to an integral number of wavelengths ($n\lambda$) in order that constructive interference can take place.

In fig. 1.5 *b* the path difference between ray AOB and ray CDB is OD–OP, as can be seen by constructing the perpendicular PD. By drawing the normal to the diffracting planes OX, and extending CD to X, if OY is parallel to PD, by geometrical similarity it is obvious that PO = DY, DO = DX, i.e. the path difference between two rays is equal to YX. Furthermore, $Y\hat{O}X = \theta$, so we can write :

$$\text{Path difference} = OX \sin \theta = 2d \sin \theta = n\lambda.$$

This is the well-known Bragg law, and it tells us that X-rays will only be diffracted by a given set of crystal planes if the wavelength of the X-rays and their angle of incidence are related by the above equation. Thus if we consider several possible sets of crystal planes, as in the diagram of fig. 1.5 *c*, diffraction will only be observed (*a*) when the angle of incidence is equal to the angle of diffraction, and (*b*) when the path difference between incident and diffracted rays is equal to an integral number of wavelengths.

If a single crystal were held in the path of an incident monochromatic X-ray beam, therefore, diffraction would not necessarily be observed, and in the study of crystal structure by X-rays it is necessary to provide a range of values either of θ or of λ to ensure that Bragg's law is satisfied. Three types of experiment can be carried out in order to do this :

(*a*) The crystal can be held in the path of a beam of X-rays which consist of a continuous range of wavelengths (so-called ' white radiation ' by analogy with white light). For a given crystal, therefore, the angle of incidence and the interplanar spacings are fixed, so that diffraction is ensured by one particular wavelength being selected out of the incident beam by a particular set of planes in order to satisfy $n\lambda = 2d \sin \theta$. The positions of the diffracted beams are usually recorded by means of a photographic plate, and this method gives what is called a ' Laue photograph ', and the technique is widely used in metallurgical research to find the orientation of the various crystal planes within a metal single crystal. The properties of such crystals can then be studied and related to their internal structure.

(*b*) Using a source of monochromatic X-rays and a single crystal specimen, one can satisfy the Bragg equation and obtain diffracted beams by varying the value of θ. This is done by oscillating the crystal to and fro about a particular axis. Under these conditions diffracted beams ' flash out ' at certain positions during the oscillation as the condition $2d \sin \theta = n\lambda$ is satisfied, and from the resultant ' oscillation photograph ' the spacing of the crystal planes perpendicular to the axis of oscillation can be calculated. Although this technique has rather less application in physical metallurgy research, it is a common method of establishing the structure of crystalline materials.

(c) Finally one can crush up the crystal into a fine powder and place this (perhaps in a fine glass tube) in the path of a monochromatic X-ray beam. In this way crystals in virtually every possible orientation will be presented to the incident beam, so that once more the correct value of θ, the angle of incidence, can be found, by the favourably orientated particles in the powder contributing to the diffracted beams. Cones of diffracted rays are produced, and by enclosing the specimen in an X-ray camera containing a narrow strip of film, a ' powder photograph ' can be obtained as illustrated in fig. 1.5 d. This method is widely used in metallurgical research, not only in the investigation of the crystal structure of metallic materials, but also in the identification of unknown substances from the characteristics of the X-ray patterns obtained.

Since a crystal consists of a regular pattern of atoms, the atomic arrangements can be described completely by specifying atom positions in some *repeating* unit, called a ' unit cell '. The cells are chosen to have a simple geometry, and if the unit is repeated indefinitely like a building-block, the structure of the whole crystal can be built up. These X-ray diffraction experiments have indicated that the arrangements of the atoms in about 80% of all metals can be classified into *three* different categories, and we will now examine these in some detail.

1.3.2. *Metal crystals*

Because the repulsions between ions are very large when they are brought close together, the ions in a metal crystal can be treated as rigid spheres, and the attractive forces between each ion and all the electrons surrounding it tend to pack the spherical ions into as small a space as possible (but at the same time keeping the same distance between each pair of adjacent spheres).

It is easy to make a model of the packing of spherical ions by packing sets of spherical objects such as ball-bearings, marbles, ping-pong balls or polystyrene spheres of uniform size. A layer of spheres (fig. 1.6) can be arranged on a flat surface so that each sphere is in contact with six other spheres in the layer ; the layer is said to form a *close-packed plane*. Note that there are three directions (at 120° to each other) along which the spheres are in contact with their neighbours —these are the *close-packed directions* in the plane.

11

We may build up a crystal model by assembling successive layers of close-packed planes one above the other, and the two regular ways in which this can be done will give rise to two crystal structures which together are possessed by 60% of the metallic elements.

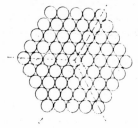

Fig. 1.6.   A close-packed planar array of spheres.

(a) The hexagonal close-packed structure

If a second close-packed layer of spheres (B) is taken and placed on top of a layer like that of fig. 1.6 so that the spheres of layer B fit into the hollows in the first layer (A), the situation will be as shown in fig. 1.7 a.

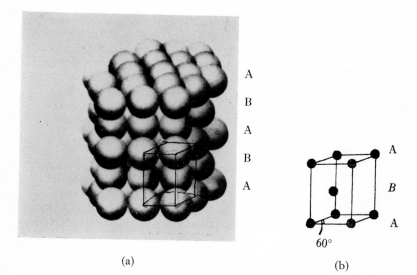

(a)   (b)

Fig. 1.7.   (a) The hexagonal close-packed structure, with one unit cell outlined. (b) The unit cell of the h.c.p. structure.

A third close-packed layer of spheres can be placed on top of (and in the hollows of) layer B, so that each sphere in the new layer is directly above a sphere in the first layer (A) (fig. 1.7 a), so that we can also

12

designate the third layer A. We can now add further layers in the sequence ABABABAB, etc., and this produces a crystal structure known as the hexagonal close-packed (or h.c.p.) structure. Each sphere touches twelve other spheres—six in the close-packed plane in which it lies and three in each of the layers above and below the plane in question.

Beryllium, magnesium, zinc and cadmium are among the metallic elements which possess this crystal structure, and the sixfold symmetry implied by its name is obvious if one considers an axis perpendicular to the close-packed planes (fig. 1.6)—where the hexagonal arrangement is clearly shown. The smallest building-block from which a hexagonal close-packed crystal can be constructed is illustrated in fig. 1.7 *b*, which represents the 'unit cell' of this structure.

(b) The face-centred cubic structure

There is an alternative sequence to the above in which the close-packed layers of spheres can be vertically stacked. When a third layer of spheres is placed on top of and in the hollows of layer B, it can be placed, not vertically above the spheres in the A layer, but with the centres of the spheres over holes in the A layer. We will call the layer in this position layer C. The fourth layer can now be laid so that its spheres are directly above those in layer A, and the fifth above those in layer B and so on, so that the pattern of stacking can be written ABCABCABC, etc., as shown in fig. 1.8 *a*. Again the spheres are arranged so that each one is in contact with twelve other spheres, and this arrangement is known as the face-centred cubic structure (f.c.c. structure). Aluminium, nickel, copper and lead are among the important metals which have this crystal structure.

Figure 1.8 *b* is an illustration of this same crystal structure, but shown in a different orientation from that shown in fig. 1.8 *a*. The second illustration demonstrates the cubic symmetry of the atomic arrangement implied by its name. A series of cubic unit cells can be seen, with spheres situated at the corners of each cube and also at the centres of each cube face. The most difficult thing to visualize is the relationship between figs. 1.8 *a* and 1.8 *b*. To show this, in fig. 1.8 *b*, a cube *corner* is 'sliced off', and one of the close-packed planes is exposed. Peeling off further layers will reveal similar planes in the ABCABC, etc. sequence. Thus the *body-diagonal* direction PQ in fig. 1.8 *c* (the diagonal from one corner of the cube to the opposite one) is the direction of stacking of the close-packed planes (shown vertically in fig. 1.8 *a*).

One further important point must be made : in a cubic unit cell, *all eight corners* are structurally identical. Although we chose to ' slice off ' corner Q in fig. 1.8 *c* in order to reveal the close-packed planes perpendicular to the diagonal direction PQ, we would have produced exactly similar results if we had ' sliced off ' any of the other corners,

13

R, S, T, U, V or W. In other words, all the body diagonals of the cube (PQ, RS, TU and VW) lie perpendicular to families of close-packed planes. In constructing an f.c.c. crystal model by assembling close-packed layers in *one* direction of stacking in the sequence ABCABCABC, etc., our final model will in fact contain *four* identical families of such planes. This arises because of the cubic symmetry

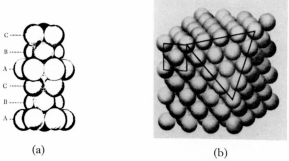

(a)   (b)

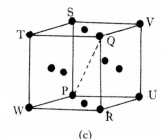

(c)

Fig. 1.8. (*a*) The formation of the f.c.c. crystal structure by the vertical stacking of close-packed planes in the sequence ABCABCABC, etc. (*b*) A cube-oriented f.c.c. crystal (one unit cell is marked) showing a family of close-packed planes perpendicular to the body-diagonal (PQ) of the unit cell (shown in (*c*)).

of the crystal; although a cube has eight corners, there are four and not eight families of close-packed planes in the structure, since by ' slicing off ' at any two diametrically opposite corners (e.g. P and Q) we expose the same family of planes.

(*c*) *The body-centred cubic structure*

The remaining metals do not solidify in a close-packed structure: about 20% of metals form crystals in which each ion is in contact with only *eight* other ions, rather than twelve, to form what is known as the *body-centred cubic* (or b.c.c.) crystal structure, which is illustrated in fig. 1.9. In the unit cell, nine spheres are arranged to form a cube in which a sphere is at each corner, and a sphere is at the centre

14

of the cube. Although there are no close-packed layers of spheres here, as in the h.c.p. or f.c.c. structures, there *are* four close-packed *directions*, formed by rows of spheres in contact along the body diagonals of the cube. This feature is important, as we shall see in the deformation behaviour of b.c.c. metals. Alkali metals (such as sodium and potassium) and transition metals (such as chromium, molybdenum and tungsten) tend to have body-centred cubic structures.

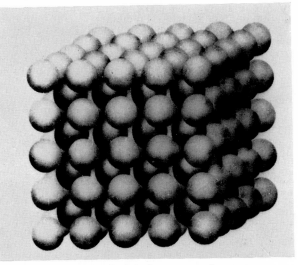

Fig. 1.9. Model of the b.c.c. crystal structure.

(d) Other crystal structures in metals

The above three simple crystal structures are confined, in the main, to what might be called the ' true metals ' and the distribution of these structures among the elements is shown on the Periodic Table inside the front cover. If a metallic bond in a crystal is partly replaced by the covalent bond, more complicated structures are usually formed. This is because the *directional* covalent bonds hold the atoms in fixed arrays of sheets or chains, and these are held together by metallic and by secondary bonds. For example, the structure of selenium and tellurium is illustrated in fig. 1.10. The structure consists of a series of spiral chains of atoms running parallel to the vertical axis of the hexagonal unit cell (cf. fig. 1.7 *b*). Each atom within the chain is strongly bonded covalently to its immediate neighbours by two bonds at an angle of approximately 100°.

The B sub-group metals, of which these are an example, demonstrate very convincingly the continuous and gradual transition from the metallic to the covalent bond—with aluminium and lead at one end of the series possessing close-packed metallic structures, and iodine at the other with a fully covalent structure. In between, as well as selenium and tellurium already mentioned, there are the other ' semi-metallic ' elements such as arsenic, antimony and bismuth,

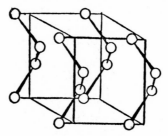

Fig. 1.10. The crystal structure of selenium and tellurium, showing the chains of atoms.

which also show a mixture of covalent and metallic bonding in their structures. We will consider later the effects of these directional bonds upon the physical properties.

(e) Polymorphism

A number of metals adopt more than one crystal structure as the temperature or the pressure is varied, and this phenomenon is known as ' polymorphism '. The best-known example of this effect, and the most important from a metallurgical point of view, is provided by iron—which is b.c.c. below 910°C (α-Fe) and above 1400°C (δ-Fe), but between 910°C and 1400°C it has an f.c.c. structure (γ-Fe).

Several metals that are h.c.p. at lower temperatures transform to b.c.c. at higher temperatures—these include lithium, beryllium, titanium and zirconium. Plutonium is particularly complex in that it has six different crystal structures between room temperature and its melting-point of 640°C.

The construction of crystal models of the common structures of metals is described in § 1.5.1 on p. 18.

1.4. *The physical properties of metallic crystals*

The physical properties of metals can be closely linked with the nature of the bonding. There is, furthermore, strong evidence that the nature of the bonding in metals is little changed in the liquid state from that in the solid state : for example, the latent heat of fusion of metals is small, and there is, furthermore, only a small increase in volume when a metal melts to a liquid.

On the other hand, the complex and ' open ' nature of the structures of the B-group metals discussed in the previous section leads to a *contraction* in volume when the solid melts. Conversely, when these metals freeze there is an expansion, and this important effect gives

16

several of these elements a technical importance. For example, *linotype metal*, which is used for printing, contains 12% antimony and 4% tin in lead which, as well as melting at a low temperature (240°C), also has the important property of expanding when it freezes, due to the formation of directional covalent bonds. This expansion enables a very sharp impression of the type-mould to be produced.

The latent heats of vaporization of metals have the same high range of values as for non-metals and ionic compounds. Since this parameter represents the energy to separate the atoms completely, it seems that the boiling-point (rather than the melting-point) marks the change to the electrons being associated with any *one* atom, instead of being ' delocalized ' in a ' gas ', or ' sea '.

We have also seen that, from the nature of the metallic bond, metallic ions tend to be packed together like close-packed spheres. This of course leads to the characteristic high *density* of metals. The free-electron theory of metals can also account for their optical opacity and for their reflectivity. In any incident light the free electrons oscillate in the alternating electric field associated with the radiation, and the energy of all light wavelengths is absorbed and the metal is opaque. The electrons then, however, re-emit the light waves as they oscillate, giving rise to the lustre and reflectivity of the metallic surface.

Tolman and Stewart provided direct evidence for the presence of free electrons in metals by accelerating a piece of metal so sharply that the free electrons were by their inertia thrown towards the rear. A detectable pulse of current was produced under these conditions, known as the ' Tolman effect ', and it was further shown that the ratio of charge to mass of the particles forming the current was the same as that of electrons.

The extremely good electrical conductivity of metals is also im- mediately explained by the drift of free electrons through the metal under the influence of an applied electric field. As this drift of electrons proceeds, they will collide elastically with the positive ions and thus lose energy so that under a given potential gradient a terminal *drift velocity* will be achieved when there is an equilibrium between the rates at which they acquire energy by accelerating in the applied field and lose it by collision with the lattice. This of course, is the origin of the rise in temperature of metals when they conduct electricity. Again, if the temperature of the metal is raised, its conductivity falls, since the increased thermal agitation of the ions further impedes the free-electron drift, so that a reduced terminal drift velocity will be observed for a given applied potential. This increase in resistivity with increasing temperature is the most apparent property that rigorously defines the metallic state.

We will defer a discussion of the deformation behaviour of metals until Chapter 4. In this chapter we have considered only the first

17

two of the three aspects of structure we initially defined. We have seen the relationship between the *electronic structure* of the metal atom and the *crystal structure* which forms. We will, in the next chapter, examine the third aspect of structure—the *grain structure* of metals.

1.5. *Experiments and problems*

1.5.1. *Construction of crystal models*

Ping-pong balls or polystyrene spheres of similar size can be used to construct models of the crystal structures encountered in metals. The individual balls should be glued together with the type of cement used for assembling model aircraft, etc.

(a) *Close-packed planes*

Assemble at least nine spheres in the form of a close-packed plane (fig. 1.6) and prepare four or five such sheets of spheres. Note (a) the number of spheres in contact with a given sphere within the sheet, (b) the number of close-packed rows of spheres per plane, and the angle between these rows.

(b) *Hexagonal close-packed structure*

Assemble three (or five) close-packed planes of spheres in the hexagonal close-packed array (fig. 1.7 a). With poster paint it is possible to colour the nine spheres that form the hexagonal unit cell (fig. 1.7 b).

Note the number of spheres in contact with a given sphere within the crystal model (this is termed the 'co-ordination number' of the structure).

(c) *Face-centred cubic structure*

Assemble four close-packed planes of spheres in the f.c.c. array (fig. 1.8 a). See if you can devise a method of marking those spheres forming an f.c.c. unit cell (fig. 1.8 c).

Note the number of spheres in contact with a given sphere within the crystal model (the co-ordination number).

(d) *Body-centred cubic structure*

Assemble nine spheres to form a b.c.c. unit cell. What is the co-ordination number of this structure?

Use these models in considering the following calculations.

1.5.2. *Crystallographic calculations*

(a) If r is the atomic radius, calculate the length of the cube edge of a unit cell of f.c.c. and of b.c.c. structure.

18

If the atomic *diameter* of copper is 2.556×10^{-10} m and that of iron is 2.481×10^{-10} m, how many unit cells are there in 1 cm³ of these elements?

(b) When unit cells are repeated to build up a crystal structure, atoms lying at the corners, edges or on the faces of the individual cells are shared with neighbouring cells (see figs. 1.7–1.9). Can you write down the effective number of atoms per cell in (a) the f.c.c. and (b) the b.c.c. structure? How many *atoms* are there per cm³ of (a) copper and (b) iron?

(c) The atomic weight of iron is 55.84. The size of the edge of the unit cell of iron has been calculated in (a) above. Calculate the density of iron with the use of Avogadro's number (6.02×10^{23} per mole).

Supplementary reading list

The Science of Engineering Materials : C. R. Tottle (Heinemann), Chapter 3.
The crystal structures of the elements are given in *Metals Reference Book* : C. J. Smithells (Butterworths).

CHAPTER 2

grain structure

2.1. *Crystallization*

THE transition from the liquid state to the solid state is known as 'crystallization', and the mechanism by which the process takes place controls the microstructure of the final product. In a liquid metal (fig. 2.1 *a*) the particles are in a state of continual random motion due to their thermal energy, and there is no ordered arrangement of atoms over long distances (i.e. distances which are many times the atomic size), so its properties will be very different from those of a solid. Atoms move about readily in this disordered array, and the liquid structure is thus able to flow freely under very small stresses, which would have no effect on the more rigid array of atoms of a solid.

In a solid metal (fig. 2.1 *b*) the atoms are arranged in a highly ordered way, and they vibrate about fixed points in the crystal. This,

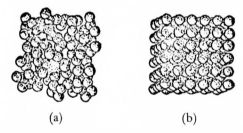

(a)   (b)

Fig. 2.1.   (*a*) The atomic array in a liquid metal.   (*b*) The atomic array in a solid metal.

we saw in Chapter 1, usually leads to a denser structure than that obtaining in the liquid, with the result that a small contraction in volume takes place when liquid metal solidifies. Let us now consider the transition in more detail.

2.1.1. *Nucleation*

During a phase transformation, such as the change from liquid to solid, it is not necessary for the entire system to transform at one jump : if it were so, phase changes would practically never occur. The process of solidification occurs by a mechanism of *nucleation* of small 'seed' crystals in the liquid, which then grow by the addition of

20

more material from the liquid. The situation is similar to the formation of liquid drops from a vapour—the formation of raindrops in a cloud is a familiar example.

Random movements of the atoms in the liquid metal (fig. 2.1 *a*) give rise to the formation of small local agglomerations or clusters of atoms having crystalline order. At temperatures *above* the melting-point of the solid these crystal ' embryos ' are always unstable and the constituent atoms are dispersed into a disordered array once again. In the molten metal such embryos are continually forming and dispersing, due to the thermal agitation of the atoms.

If the liquid metal is cooled below its melting-point, i.e. it is *supercooled*, then some of these embryo crystals may become stable and act as *nuclei* for the growth of grains of the solid. Whether or not an embryo can form a stable nucleus depends upon the degree of supercooling of the liquid and upon the size of the embryo, in the manner illustrated in the graph shown in fig. 2.2. This figure shows that the size (r_0) of the smallest self-sustaining nucleus is inversely proportional to the degree of supercooling (ΔT).

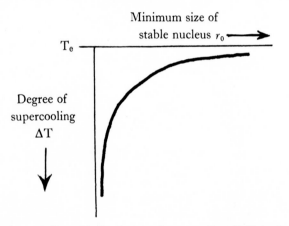

Fig. 2.2. The variation with the degree of supercooling (ΔT) in the size of the smallest stable nucleus (r_0). T_e is the melting-point.

Thus for small degrees of supercooling (ΔT small), r_0 is large, and there is only a low probability that a large embryo will be formed in the liquid in a given time by random thermal motion of the atoms. There is thus likely to be only a low number of successful nuclei per unit volume of liquid. For high degrees of supercooling (ΔT large) r_0 is small, and the probability of forming such a nucleus is now very high, so that a high number of successful nuclei per unit volume of liquid will be observed. The implications of these effects upon the resultant structure will be discussed below.

2.1.2. *Growth of nuclei*

Once stable nuclei are formed in the liquid, they grow at the expense of the surrounding liquid until the whole volume is solid. When most commercial metal and alloys are cast and allowed to freeze in a mould, the nuclei are observed to grow more rapidly along certain crystallographic directions. For example, in f.c.c. and b.c.c. crystals, the directions of the cube edge of the unit cell are directions of rapid growth ; this causes spike-shaped crystals to develop and other arms may branch out sideways from the primary spikes like branches from the trunk of a tree. This results in the formation of crystals with a three-dimensional array of branches, known as

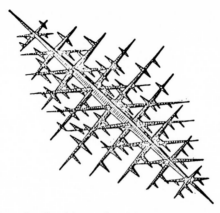

Fig. 2.3.   Drawing of a dendrite : the side-branches grow at various angles to the main stem to form a three-dimensional skeletal structure.

dendrites, as shown in fig. 2.3.   Section 2.5.1 (p. 37) describes an experiment to demonstrate the formation of dendrites.

2.1.3. *The grain structure of metals*

Figure 2.4 shows schematically how the structure of a solid metal develops by solidification of the liquid.   Dendrites grow outwards

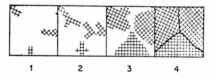

Fig. 2.4.   Schematic view of stages in the freezing of a liquid to a polycrystal by the nucleation and growth of dendrites.

from each crystal nucleus until they meet other dendrites from nearby nuclei : growth then halts and the remaining liquid freezes in the gaps between the dendrite arms. Each original nucleus thus produces a grain of its own, separated from the neighbouring grains by a *grain boundary*—which is a narrow transition region in which the atoms adjust themselves from the arrangement within one grain to that in the other orientation.

The position of the grain boundaries is determined by where the separately growing dendrites happen to meet. If there is a high number of nuclei growing in the liquid, then the resulting polycrystalline solid will be of fine grain size. If the number of nuclei formed on freezing is small, then a larger grain size will form ; if (by controlled solidification) only one nucleus is allowed to grow, then obviously a *single crystal* solid will result. In many fields of research in physical metallurgy and solid-state physics the use of single crystal specimens is extremely important, and some applications will be discussed in later chapters.

When liquid metal is poured into a cool mould, the layer of liquid next to the wall of the mould is cooled very rapidly. This gives rise to a very large local supercooling with the result that very many small nuclei of the solid are formed upon the mould wall, which grow to produce the very fine-grained layer of crystals at the surface of the casting, known as the ' chilled layer '. These grains in the chilled layer have similar dimensions along all axes and are said to be equiaxed crystals.

As the shell of solid metal thickens, the crystals on the inside of the chilled layer begin to grow inwards to form long columnar crystals whose axis is parallel to the direction of heat flow. Crystals with orientations favourable for rapid growth grow faster than their neighbours across the mould until they meet the crystals growing from the opposite wall.

It is of great technical importance to produce uniform, equi-axed crystals in large castings, for this structure is associated with good strength and other mechanical properties. This may be done by adding a *nucleating agent* to the metal, whose action is to form tiny seed crystals in the molten liquid when it is in the mould, so that the whole structure freezes from many nucleating centres to form fine-grained solid. Section 2.5.2 (p. 39) describes an experiment to demonstrate the structure of castings.

2.2. *Metallography*

The sizes and shapes of grains in a material can vary over a wide range, depending on how the metal has been formed. Grains as small as 0·01 mm are common, though as-cast metal usually has coarser grain size in the range 0·1–10 mm. Occasionally these can be seen with the naked eye—for example, the large grains of zinc on the surface of a new sheet of galvanized steel are often very clear (on a new dustbin for example), and again one sometimes sees the individual grains on a cast brass door-knob that has been polished by handling and etched by perspiration from the hands. More usually, however, it is necessary to prepare a *section* from a metallic object in order to study the size, shape and distribution of crystals within it, which is usually called a *metallographic examination.*

In preparing a metallographic section, precautions are taken at every stage to ensure that the method of preparation does not itself alter the microstructure originally present. To this end, accidental distortion or heating of the specimen must always be avoided, and although the section for study may have to be cut from the bulk by a process of milling or sawing or by the use of an abrasive cutting wheel, precautions are always taken to prevent its temperature from rising by providing ample cooling and lubrication during this and all later stages.

Gross distortions from the cutting process must next be eliminated, which can be done by first filing the surface and then by grinding it with successively finer abrasives. A plane, smooth surface can result from this process, which usually involves abrading the surface with firm even strokes upon a series of emery papers or silicon carbide papers of progressively finer particle size supported on a smooth surface such as a glass plate. If the grains are coarse enough to be seen with the naked eye, one can at this stage prepare the surface for *macroscopic examination.*

This involves etching the surface of the specimen, usually in a dilute acid, by immersing it in or swabbing it with a suitable reagent until the individual grains are revealed. Due to the different rates of chemical attack along different planes in a crystal, when the surface is etched, crystallographic terraces are formed upon each grain as shown in fig. 2.5. These terraces reflect the light in directions which vary

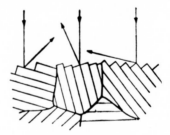

Fig. 2.5. The effect of an etchant upon the surface of a polycrystalline metal.

with the orientation of the grain, so that some crystals appear light and some dark to the eye. The macrostructure of a piece of cast metal which has been prepared in this way is shown in fig. 2.6.

In this specimen the grain size is of the order of a few centimetres, so that there is no difficulty in resolving the structure without magnification. In most familiar metallic objects, however, the grain size is too fine to be discerned without the use of a microscope.

Fig. 2.6. The macrostructure of a cast metal. Large equiaxed grains have formed at the centre of the ingot. (Courtesy R. T. Southin.)

2.2.1. *The metallurgical microscope*

The principle of the metallurgical microscope is illustrated in fig. 2.7 *a*. A horizontal beam of light enters an aperture in the microscope tube, where it strikes a glass slip inclined at an angle of 45°. About 10% of the light is reflected down through the objective lens, where it strikes the specimen surface and may be reflected back through the objective lens to form the primary image, which is then magnified further with the eyepiece. Such an arrangement with the best objectives gives magnifications of over 2000 ×, with a resolution permitting details of the specimen to be observed which are separated by about 1 μm.

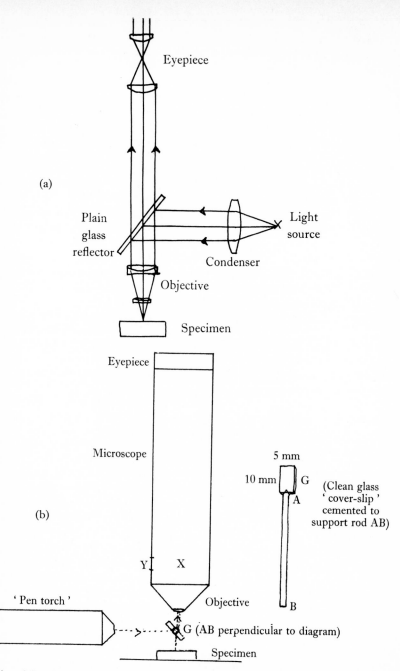

(a)

Eyepiece

Plain glass reflector

Light source

Condenser

Objective

Specimen

(b)

Eyepiece

Microscope

5 mm

10 mm   G

(Clean glass 'cover-slip' cemented to support rod AB)

A

Y   X

'Pen torch'   Objective   B

G (AB perpendicular to diagram)

Specimen

Fig. 2.7. (*a*) The principle of the metallurgical microscope. (*b*) The adaptation of a biological microscope for use by reflected light. (If microscope is expendable, tube may be bored at X and Y so that illuminating system is above the objective lens.)

The illumination system illustrated in fig. 2.7 *a* is essential when working at high magnifications. At low magnifications, however, because the distance between the specimen surface and the objective lens is greater, it is possible to adapt an ordinary biological microscope to examine a metal surface. A glass cover-slip, about 3–4 mm in width is mounted so that it can be rotated between the specimen and the objective lens as shown in fig. 2.7 *b*, and a pen-light provides a suitable source of light. Total magnifications of up to 200 × can be obtained by means of such an arrangement.

Specimen preparation for microscopy is much more critical than for macro-studies, and very careful polishing must be carried out in order to remove all trace of the distortion produced in the initial grinding stage, since the fine emery scratches obscure the details of the metal structure. Polishing to a mirror finish is often done in two stages, with a coarse and a fine polishing agent respectively. The specimen is usually held against a horizontal rotating wheel covered with a short-pile cloth fed with a suspension or cream of the polishing agent. Wheel speeds are normally from 100–300 r.p.m., and polishing agents include suspensions of very fine particles (1 μm particle size) of magnesium oxide or aluminium oxide, although commercially produced domestic ' metal polish ' can sometimes be used with success. Another way of polishing a specimen is to make it the anode in a suitable electrolyte, when, if the current density is correct, a bright scratch-free surface can be produced—this technique is known as electrolytic polishing.

A much lighter etching treatment is applied for microscopical examination than for macro-studies. With some etching reagents and very short etching times, metal is dissolved only at the grain boundaries, giving rise to shallow grooves there, which are seen as a network of dark lines (figs. 2.8 *a* and *b*) under the microscope. Figure 2.8 *b* shows the microstructure of a commercially pure iron which is typical of many polycrystalline metals : as grain boundaries are regions where the atom arrangement is distorted, they have a high energy, and hence there is a tendency for the boundaries to assume the smallest possible surface area. This effect is found with soap bubbles : a foam of soap bubbles assumes shapes that make the total surface area a minimum. Along an edge where three bubbles meet, the surface tension of each being the same, the boundaries will have angles of 120° between them, and metal grains will also tend to give angles of 120°, although a micro-section will not show this if the line of intersection of the grain-boundary surfaces is not perpendicular to the plane of the section.

With some metals, the etching may produce an effect similar to that sketched in fig. 2.5, although on a much finer scale than that encountered in macro-etching. This results in different rates of attack and roughening of the surfaces of the variously oriented grains, so that the surface reflectivities of the grains are changed by different amounts. A similar effect occurs if the material contains more than one *phase*— i.e. there are regions of different crystal structure and composition (see Chapter 3). Many important *alloys* contain such features, and

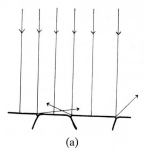

(a)

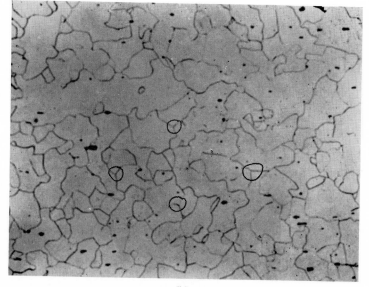

(b)

Fig. 2.8. (a) The action of a grain-boundary etchant. (b) The microstructure of pure iron, magnification × 200. Some grain corners which tend to have 120° angles are marked.

the different phases are revealed under the microscope by virtue of their different rates of attack by an etching reagent.

When some metals of f.c.c. crystal structure are examined microscopically, as well as seeing the irregular network of grain boundaries, parallel-sided regions known as *annealing twins* are apparent within many of the grains (fig. 2.9). These bands are regions in which the crystal structure has a new orientation, and they come into existence during the growth of the crystals. In the twinned part of the crystal (fig. 2.10) the atom arrangement is related to that in the 'parent' crystal by a mirror-image relationship across the twin interfaces. A simple way of regarding annealing-twin formation is to consider the crystal forming by the assembly of close-packed planes in the sequence ABCABCABC, etc. as described on p. 13. When an annealing twin forms, this can be regarded as arising from a change in the stacking sequence to give a sequence which is a mirror-image of the original, as illustrated in fig. 2.10. These twin bands are a very common feature of the microstructures of many important materials, such as copper and nickel and many of their alloys.

The upper limit of magnification of the optical microscope is inadequate to resolve many structural features which are important

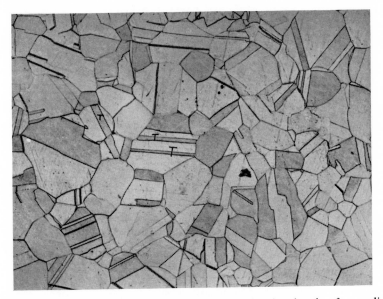

Fig. 2.9. The microstructure of pure copper, showing bands of annealing twins in many grains (e.g. at T).

in metallurgical specimens. In recent years the electron microscope has come into common use by metallographers, and we will next consider several techniques which permit metals to be studied at very high magnifications.

2.3. *The electron microscope*

The principle of the electron microscope is exactly analogous to that of the optical microscope, with a fine beam of electrons replacing the light beam, and the lenses consisting of carefully designed electro-magnets which focus the electron beam to produce an image on a fluorescent screen. The whole system is enclosed in a chamber maintained under very high vacuum conditions, there being an 'air-lock' device to enable specimens to be inserted and removed from the

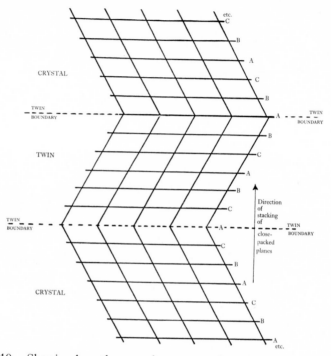

Fig. 2.10.   Showing how the crystal structure of a twin is the mirror-image of the parent crystal.

instrument without losing the vacuum. A schematic diagram of a typical electron microscope is given in fig. 2.11. The magnetic lens system is analogous to that in an optical microscope, but the focal length of the lenses is controlled by regulating the current through the coils of the lenses, and magnified images of up to $100\,000\times$ may be obtained on the viewing screen.

Just as visible light can be regarded as having the characteristics of both waves and particles, so electrons can be looked upon both as particles and as waves. This wave-like character was most clearly demonstrated by de Broglie's experiments showing the diffraction of electrons by crystals. The inter-relationship can be seen by equating the well-known Einstein equation for energy and mass:

$$E = mc^2,$$

30

(where E = energy, m = mass, c = the speed of light) with the suggestion by Planck that energy in the form of radiation can only be omitted in units or quanta of magnitude :

$$E = hv$$

(where v = frequency of the radiation and h = Planck's constant of action), i.e. $hv = mc^2$.

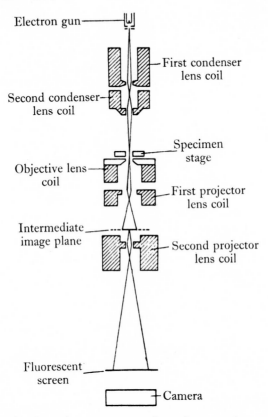

Electron gun

First condenser lens coil

Second condenser lens coil

Objective lens coil

Specimen stage

First projector lens coil

Intermediate image plane

Second projector lens coil

Fluorescent screen

Camera

Fig. 2.11. A schematic arrangement of an electron microscope system.

Since velocity is equal to the product of wavelength (λ) and frequency, i.e.

$$c = v\lambda,$$

we can write :

$$m = \frac{hv}{c^2} = \frac{h}{\lambda c}$$

or

$$\lambda = \frac{h}{mc} ;$$

31

so that any particle of mass m and velocity v would have a wavelength given by $\lambda = h/m.v$.

The wavelength of the electrons in the electron microscope is dependent upon the accelerating voltage (V) applied to the electron gun, and is given approximately by

$$\lambda = \sqrt{\left(\frac{150}{V}\right)} . 10^{-10} \text{ m}.$$

Normal operating voltages are of the order 100 kV, so that λ may be about 3.5×10^{-12} m.

In order to form an image in any optical system, the wavelength has to be small compared to the size of the features which it is desired to distinguish or resolve. The optical microscope is limited by the wavelength of light $(5 \times 10^{-5}$ m$)$ to objects of size about 10^{-6} m. The much shorter wavelength possessed by electrons gives rise to an enormous improvement on this, and modern electron microscopes can image structural details down to a size of about 10^{-9} m, limited only by the present state of magnetic lens design.

2.3.1. *Electron metallography*

With the advent of this instrument, metallographers were confronted with a difficult problem since the specimen is viewed by focusing electrons transmitted through the specimen, and metals readily absorb electrons. Initially the problem was solved by the development of replica techniques. For example, an imprint of the differences in height on a polished and etched metal surface may be obtained by adding a dilute solution of a plastic material in a volatile

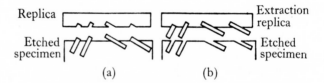

Replica   Extraction replica

Etched specimen   Etched specimen

(a)   (b)

Fig. 2.12. (*a*) Principle of surface replication. (*b*) Principle of the extraction replica.

solvent (rather like ' nail varnish ') to the metal surface. When the solvent evaporates a thin film of plastic is left, the upper surface of which is flat, while the lower follows the contours of the etched metal (fig. 2.12 *a*). More recently, improved replicas have been made by evaporating a thin film of carbon onto the metal surface, while in an evacuated chamber.

The replica is stripped from the metal surface, often by chemically attacking the underlying metal, and a piece of it about 2–3 mm in diameter is placed onto a small copper supporting grid which is then inserted in the electron microscope. The contrast in the image is obtained by the differences in the scattering of the electrons as they pass through the thin and thick regions of the replica, which correspond to the raised and lowered parts of the metal surface.

In the case of specimens containing small particles of a different kind of crystal, as in some alloys (see Chapter 3), it may be possible to retain in the replica, when it is stripped from the specimen, the actual particles which originally lay in the polished surface of the specimen (fig. 2.12 b). The latter are known as ' extraction replicas ' and they permit the direct examination of such particles in metals and alloys. Extraction replicas have proved particularly useful in the study of steels, where it has been possible to obtain a clear understanding of the structural changes brought about by the heat treatment of these materials. The method has also been applied with marked success to systems based on uranium and to nickel-based alloys used for high temperature applications.

Although the study of surface replicas is still made in metallurgical research, a much wider use is now made of very thin metal samples. Great skill is required to produce metal foils whose structure is characteristic of the metal in bulk, yet whose dimensions are of the order of 2–3 mm in diameter and 10^{-4} mm in thickness (that is smaller than the wavelength of light). Foils are usually prepared by chemical or electrolytic thinning of the material, so that it cannot suffer mechanical damage.

The contrast observed in the electron microscope image from crystalline specimens (i.e. in thin metal foils or in extraction replicas) varies with the orientation between the specimen and the beam, and also with the specimen thickness, and it arises from the interactions of the electron waves with the atomic arrays in the specimen. Elaborate theories of the nature of the electron–atom interactions have been developed to account for the contrast effects observed.

As well as producing real magnified images, modern electron microscopes have facilities to enable *electron diffraction* experiments to be carried out : this technique is used with any crystalline specimen. The process is exactly analogous to X-ray diffraction as described in § 1.3.1, and from the interference patterns observed the spacing of sets of crystal planes can be calculated. In this way unknown structures can be identified, and the *orientation* of crystals established with respect to the direction of the electron beam.

2.3.2. *The high-voltage electron microscope*

The most severe limitation of the conventional electron microscope in the study of thin metal foils is the necessity of having extremely thin specimens, in order that they are transparent to the electron beam. It is not valid to attempt to observe directly any phase change or deformation process while such a thin specimen is actually in the microscope, since the proximity of the surface causes the material to behave in a way uncharacteristic of the bulk solid. Again, the transparency of a thin metal foil to the electron beam is proportional to the atomic number of the element, so that it is difficult to examine metals such as uranium in the ordinary electron microscope.

In recent years electron microscopes have been developed with electron-gun voltages of up to 1000 kV. These are now commercially available, but are major items of research equipment, and require a whole building to house them. The high-energy electrons are much more penetrating than those used in conventional electron microscopes, so that foils of heavy metals can be easily examined, and in the case of light metals such as aluminium, specimen thickness can be increased by an order of magnitude. The possibility of performing sophisticated 'in-microscope' experiments now presents itself, with the surface of the foils no longer dominating the behaviour of the material, so that much significant research into solid-state processes can be expected in the coming decade.

2.3.3. *The scanning electron microscope*

Optical metallography had its beginnings in the nineteenth century, particularly in the work of Henry Clifton Sorby, who studied a range of metallurgical structures by reflection microscopy from polished and etched sections. Prior to this work, attempts had been made to examine the structure of metals by studying fracture surfaces, but this met with little success since these are very irregular, and high power microscopes have an extremely small depth of focus, so that virtually only a planar field, perpendicular to the optic axis, can be examined.

The electron microscope has a comparatively large depth of focus, and by preparing replicas from fracture surfaces of metals the instrument has been able to make some important contributions to our understanding of the fracture processes in metals.

However, a modified electron microscope has recently been devised which has enabled topographic information to be obtained directly from a metal specimen. A relatively large specimen (e.g. a 10 mm cube) can be used, placed within the column of the microscope, and then its surface is scanned with an electron beam focused to a diameter of $2–3 \times 10^{-8}$ m. This beam scans a small area on the specimen in the same way that a television screen is scanned, and the electrons scattered from the surface of the specimen are collected by a detector, and the signal from this detector is used to modulate the brightness of a beam scanning the fluorescent screen of a cathode-ray tube in synchronism with the electron beam scanning the specimen. A television 'picture' of the small area of the surface is thus produced.

The scanning electron microscope can be used for examining metal fracture surfaces at high magnifications (i.e. up to $50\,000 \times$), and also with high depth of field, and an example is given in fig. 6.9 (see p. 117). It can also be used to study wear processes on abraded surfaces, as well as the early stages of oxidation and film formation upon metallic surfaces.

2.3.4. *The electron probe microanalyser*

If an ordinary metallographic specimen is irradiated *in vacuo* with a finely focused beam of electrons, those atoms in the specimen surface immediately under the beam will tend to have their inner shell electrons displaced by the bombarding electron beam. Outer electrons now change their level to replace the displaced inner ones, and liberate X-rays as they do so. The wavelength of these emitted X-rays will be

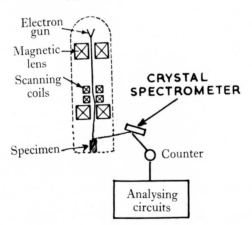

Fig. 2.13. Schematic diagram of the electron probe microanalyser.

characteristic of the element whose atoms are excited in the sample, and if the X-rays are recorded and analysed, it is possible to identify and to analyse quantitatively elements in the regions of the surface of the specimen.

This technique is known as X-ray microanalysis, and the principle of operation is illustrated in fig. 2.13. The specimen is mounted on a table which can be rotated to bring the specimen under an optical microscope so that its structure can be correlated with the other observations.

The X-rays excited emerge from the column and enter an X-ray spectrometer which enables the individual wavelengths of the characteristic spectra to be identified, and hence the elements present in the specimen can be recognized. For quantitative determinations, these X-ray peaks can be compared with similar peaks produced by pure standard samples of each element. The most modern instruments are capable of detecting and measuring the X-ray spectra of elements of atomic number down to and including beryllium.

The technique is of prime importance in the study of small-scale variations in compositions in alloys, such as are encountered in studies of precipitates and inclusions in alloys, and of corrosion effects—particularly the formation of oxide and other scales on metals.

2.4. *The field-ion microscope*

Metals may be examined at magnifications in excess of $1\,000\,000 \times$ by means of the field-ion microscope, which permits the individual atoms of the crystal to be resolved. At present it has only been

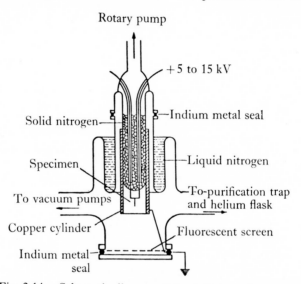

Fig. 2.14. Schematic diagram of the field-ion microscope.

extensively used in the examination of metals of high melting-point such as tungsten and molybdenum.

The experimental arrangement is essentially quite simple, and a sectional diagram of the apparatus is shown in fig. 2.14. The specimen is in the form of a needle, the point of which is hemispherical with a diameter of a few 10^{-8} m. It is mounted in a vacuum chamber on an electrode which is cooled by liquid hydrogen or other coolant, in order to minimize the thermal vibrations of its ions.

A trace of gas (usually He or Ne) is allowed to leak into the chamber, and a fluorescent screen is located a few centimetres away from the specimen tip. The specimen is held at a *positive* potential of about 5 kV relative to the screen : this causes the free electrons in the metal to be pulled inwards slightly, so that the positively charged metal ions on the tip surface are partly exposed—this effect is particularly marked where the edges of close-packed planes meet the surface of the hemispherical tip.

The image of these partially exposed ions is then carried to the screen by the gas atoms, because such an atom in the vicinity of the surface gives up an electron to the metal and hence becomes a positive ion. This gas ion then accelerates down the radial electric field lines to the screen, where it causes a scintillation. The importance of the radial field is that the pattern on the screen reproduces the pattern of ions on the tip. The magnification of this image is the ratio of the tip–screen distance to the radius of the tip and is thus of the order $1\,000\,000 \times$ for a tip–screen distance of 10 cm.

If the field strength is too low, the imaging gas will not be ionized. If, on the other hand, the field strength is too large the metal ions themselves are drawn out of the tip and the image is unstable. By careful control of the electric field it is possible to strip atoms from the tip layer by layer, with great precision, and thus to study the variation of the atomic structure with depth.

Figure 2.15 is a field-ion micrograph of iridium. The surface shows crystalline facets where resolvable crystal planes happen to be parallel to the surface. In this illustration a grain boundary is present in the specimen, and the radical change in structure across it can be seen. Individual atoms are resolved in the image as bright spots, and the width of the boundary between the two crystals can be seen to be only one to three atomic diameters.

2.5. *Experiments on solidification*

2.5.1. *Formation of dendrites*

There are a number of transparent organic materials which solidify in an analogous way to metals, and these can easily be studied using a normal biological microscope with transmitted light. A small quantity of *cyclohexanol* (m.p. 23°C) is required for this experiment.

Sufficient cyclohexanol is melted on a glass microscope slide to form a droplet about 1 cm in diameter, by holding the slide over a low bunsen flame. A glass cover-slide is then placed over the molten droplet, and while the cyclohexanol is still molten it is placed under the microscope.

As the droplet cools, the growth of the dendritic crystals can be followed under the microscope. It is possible to repeat the experiment using different cooling rates, using (for example) a small soldering iron as a heat source to decrease the rate of solidification or a piece of ice (held under the glass slide) to accelerate the process.

2.5.2. *The macrostructure of castings*

The distribution of columnar and equi-axed grains in a casting may be readily determined if *stearic acid* is solidified and examined, as this substance provides a convenient analogue of a metal.

4

Fig. 2.15. Field-ion micrograph of iridium showing a grain boundary. (Courtesy B. Ralph and T. F. Page.)

A glass tube of approximately 3 cm diameter and 15 cm length will form a suitable mould if one end is securely sealed with a rubber stopper. The stearic acid is first melted in a beaker by warming over a low bunsen flame, and then poured into the glass tube (held vertically) until it is full.

After the stearic acid is completely solid, the cast bar may be extracted by pushing it out of the tube and its grain structure subsequently examined by splitting it open longitudinally. This can be effected by making a fine saw cut a few millimetres deep along the length of the bar and across the end faces, and then splitting it in the plane of the cut by placing (say) a metal ruler along the bar in the cut and then tapping it sharply. The macrostructure of the casting will be revealed on the fracture surfaces.

The experiment may be repeated to explore the effect of increased cooling rate upon the ingot structure. If the glass tube is immersed in a beaker containing running cold water, and then the melted stearic acid is added, a change in macrostructure will be observed on the fracture surface. Note particularly the length of the columnar grains in each of the two castings. How may the differences be accounted for ? Compare your results with the illustration of fig. 2.6.

Supplementary reading list
Modern Metallography : K. H. G. Ashbee and R. E. Smallman (Pergamon).

CHAPTER 3

alloy structures

3.1. *Introduction*

An alloy is essentially a mixture of two or more elements, the principal component being a metallic element (the ' parent metal '), so that the resultant mixture exhibits metallic properties. A wide variety of mechanical and physical properties may be obtained by alloying, so that alloys, rather than pure metals, are of the greatest importance for engineering use.

Alloys can usually be prepared by melting a known mass of the parent metal in a crucible, and then dropping in and dissolving weighed amounts of the solid alloy additions (the ' solute elements '). The liquid alloy is then cast into a mould, where it freezes to a solid, and the resulting structure will depend upon whether the various types of atom are chemically indifferent to one another, or not. If they are indifferent to one another, the atoms will crystallize as a single set of crystals, since all the atoms will behave as if they belonged to the same species. A single-phase *solid solution* is said to form, and its microstructure is often indistinguishable from that of a pure metal (e.g. fig. 2.8). There may alternatively be a tendency for the elements to crystallize separately to form distinct and different crystals joined at mutual grain boundaries. Such a structure is known as a *phase mixture*, and it can usually be distinguished from a single-phase solid solution by means of microscopical examination.

3.2. *Solid solutions or phase mixtures?*

In a solid solution the crystal structure is the same as that of the parent metal—the atoms of the solute element are distributed throughout each crystal, and a range of composition is possible. The solution may be formed in two ways : (*a*) in *substitutional* solid solutions the atoms share a single common array of atomic sites, and this is shown diagrammatically in fig. 3.1 *a*. A copper crystal can dissolve up to approximately 35% zinc atoms in this way to form the familiar alloy known as brass ;

(*b*) in *interstitial* solid solutions the atoms of the solute element are small enough to fit into the spaces *between* the parent metal atoms, as illustrated in fig. 3.1 *b*.

Because of the atom size limitation involved, interstitial solid solutions are less common than substitutional solutions, although carbon atoms can dissolve in iron crystals in this way in steel.

A few pairs of metals are completely miscible in the solid state and are said to form a ' continuous solid solution ' ; copper and nickel form an important system that behaves in this way—the solution containing 70% nickel and 30% copper is known as Monel metal and has excellent resistance to corrosion which enables it to be used in

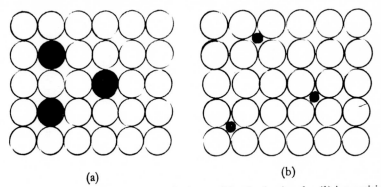

(a)   (b)

Fig. 3.1.   The formation of solid solutions : (a) substitutional ; (b) interstitial.

chemical engineering and oil refinery plants.   More usually a *limited* solid solubility is encountered, so that when larger amounts of alloying metal are added in excess of this limit a ' two-phase ' microstructure is formed by the appearance of crystals of a different structure.   Figure 3.2 is an example of an alloy of this type whose structure can be

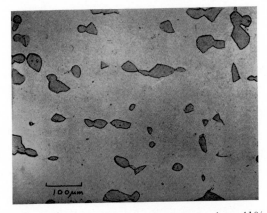

Fig. 3.2.   The microstructure of a two-phase magnesium–$4\frac{1}{2}$% cerium alloy.

interpreted in this way : the first-formed or ' primary ' phase is saturated with solute element, so that new crystals of the ' secondary ' phase must form which will in general consist of crystals of a different structure containing a higher proportion of solute atoms to parent

41

metal atoms than in the primary phase. As more and more alloying element is added to the alloy, more of the secondary phase will be produced and less of the primary phase, until eventually only the secondary phase is present, and once more the microstructure will be the same as for a pure metal. Further additions of the alloying element may produce another phase, and the sequence of changes may be repeated.

The *extent* of substitutional solid solubility of one metal in another is determined by several factors, and Hume-Rothery has generalized the experimental facts and has enabled predictions to be made by stating three broad principles:

(i) *The atomic size factor.* When the difference between the atomic diameters of the two metals is more than 14%, solid solubility is likely to be small. The atomic size itself can be deduced from X-ray diffraction measurements of the closest distance of approach of atoms in crystals of the pure element.

(ii) *The electrochemical factor.* When one of the elements is considerably more electropositive than the other, there is a greater tendency to form compounds rather than solid solutions, and the smaller is the solubility. This factor can thus be assessed from the relative position of the elements in the Periodic Table.

(iii) *The relative valency factor.* This factor appears to be less generally applicable than the previous ones, but is valid for alloys of copper, silver or gold with metals of higher valency, and it states that a metal of lower valency is more likely to dissolve one of higher valency than vice versa. For example, it is known that copper has a greater solubility for tin than tin has for copper, in accordance with Hume-Rothery's principle.

3.3. *Some simple phase diagrams*

The composition and the distribution of the phases present in a given alloy are usually most easily understood by reference to the relevant *phase diagram*. In the case of a binary alloy (i.e. a system with two components), the phase diagram consists of a two-dimensional plot of temperature versus composition, which marks out the composition limits of the phases as functions of temperature. The scale of composition of binary alloys can be expressed in several ways on phase diagrams. For an alloy of element A and element B the two commonest scales are (*a*) percentage by weight, which is given by:

$$\frac{\text{weight of B} \times 100}{\text{weight of (A + B)}}$$

in the sample (written wt. %), and (*b*) atomic per cent, which is

$$\frac{\text{number of atoms of B} \times 100}{\text{total number of atoms}}$$

in the sample. By reference to such diagrams, metallurgists can often predict the composition of those alloys which are likely to have useful properties, as well as predicting whether various thermal treatments will have advantageous or deleterious effects upon these properties.

We will therefore introduce some simple examples of phase diagrams, which we can correlate with some microstructures, and some references for further reading on this topic are given at the end of the chapter.

3.3.1. *Complete mutual solid solubility*

As stated in § 3.2, copper and nickel form a continuous solid solution, and the phase diagram for this system is shown in fig. 3.3.

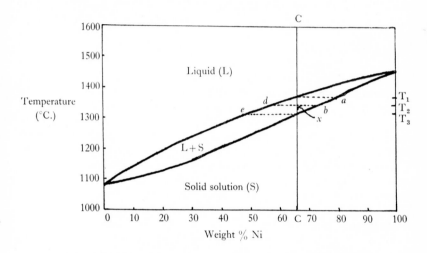

Fig. 3.3.   The phase diagram of the copper–nickel system.

The horizontal scale shows the variation in composition in weight per cent of nickel and the vertical scale is the temperature in °C. The diagram is divided into three ' phase fields ' by two lines—the upper phase boundary line is known as the *liquidus* and the lower line as the *solidus*.   At temperatures above the liquidus line, alloys of all compositions from pure copper to pure nickel will be liquid, while at temperatures below the solidus line, all alloys are in the solid state. It will be apparent that, unlike pure metals, alloys freeze over a *range* of temperature whose magnitude depends upon the composition of the alloy and is equal to the vertical separation of the liquidus and solidus at a given composition.

In working from a phase diagram, the beginner should always first consider some *specific composition* of alloy (say *C* in fig. 3.3) and study its behaviour with respect to change in temperature by constructing a

43

vertical line at that composition (see fig. 3.3). Considering a slow progressive decrease in temperature, at temperatures above T_1 the liquid phase is stable, but at T_1 solidification commences, and the two-phase field (marked L + S in fig. 3.3) of the diagram is entered. In any two-phase field of a phase diagram, the *compositions* of the two phases co-existing at a given temperature are obtained by drawing a horizontal (or *isothermal*) line.   The required compositions are given by the intersections of the isotherm with the phase boundary lines.

For example, in the present case, the isotherms are shown as dotted lines in fig. 3.3, and at temperature T_1, liquid of composition C starts to freeze by depositing crystal nuclei of solid solution of composition a, obtained by drawing the isothermal line at temperature T_1 in the two-phase field.   As the temperature continues to fall, the loss of this nickel-rich solid causes the liquid's composition to become richer in copper, which it does by following the line of the liquidus, so that when temperature T_2 is reached, the composition of the liquid (given by the new isotherm) is now seen to be d.   The growing crystals (normally in the form of dendrites, fig. 2.3) remain homogeneous, provided that the temperature is not falling too quickly, and their composition follows the line of the solidus as they cool until, at temperature T_2, their composition is given by b.   This crystal growth occurs by the deposition of layers of atoms which are richer in copper content, but atomic migration takes place within each dendrite between the new layers and the original nucleus, to enable the composition to adjust itself to b.   This can occur because in the solid state, although the atoms tend to vibrate about *fixed* points, there is sufficient thermal energy available as the melting-point is approached to enable atoms to leave their sites from time to time and move to neighbouring sites, so that a process known as *solid-state diffusion* occurs, whereby atoms can migrate within the crystal.

The dendrites we are considering will be at a temperature very close to their melting-point, so that this diffusion process can continue to allow the dendrites to adjust their composition to follow the line of the solidus as the temperature continues to fall slowly—the remaining ' mother liquor ' following the line of the liquidus.   When temperature T_3 is reached, the last liquid (of composition C) freezes, and the accompanying solid-state diffusion brings the now completely frozen solid to the composition C once again.   The solidified alloy is now (below T_3) in a single-phase field once more, and is thus stable at all lower temperatures.

In summary, therefore, we see that in the slow solidification of a solid solution alloy, although we started with a liquid alloy of composition C and finished with a set of solid crystals of composition C, the process was more complicated than in the simple freezing of a pure metal.   The initial nuclei were seen to have a different composition from the liquid in which they formed, and both the liquid phase and

44

the solid phase progressively changed their composition during the process of solidification.

The lever rule

In the temperature range T_1–T_3, when the two phases (S + L) were present, the construction of isothermal lines was shown (fig. 3.3) to give the *composition* of the two phases which were in equilibrium. This same construction also determines how *much* of each phase is present at a given temperature, for a given alloy. Consider again an alloy of composition C (% Ni) ; if at temperature T_2 the fraction of the alloy which is liquid if f_L, and the fraction of the alloy which is solid is f_S, then

$$f_L + f_S = 1.$$

If the concentration of nickel in the liquid phase = d and the concentration of nickel in the solid phase = b, then

$$b.f_S + d.f_L = C,$$

but

$$f_S = 1 - f_L,$$

so

$$C = b - b.f_L + d.f_L,$$

i.e.

$$f_L = \frac{b - C}{b - d} \quad \text{and} \quad f_S = \frac{C - d}{b - d}$$

at this temperature. These relationships are known as the 'lever rule' because an isothermal 'tie-line' within a two-phase region may be considered as a lever of length bd whose fulcrum is at the point x (fig. 3.3) where the line representing the composition (C) of the alloy intersects the isothermal line. The fraction of a phase having a composition indicated by *one* end of the lever is equal to the ratio of the length of the lever on the *far side* of the fulcrum to the total lever length.

This construction is applicable to *all* two-phase regions on phase diagrams—e.g. to the diagrams to be discussed below which contain regions where two solid phases co-exist. The lever rule is of great value to the metallographer in assessing the approximate composition of alloys from the relative proportion of the phases present that are observed in the microscope.

Problem 1 (*a*) on p. 62 relates to the system illustrated in fig. 3.3.

Non-equilibrium conditions

If a copper–nickel alloy of composition C is cooled at a fast rate, the solid-state diffusion processes described above may require too long a time to complete, so that the composition changes cannot conform to the solidus. Diffusion in a liquid can, however, take place more readily because the movement of the atoms is much less restricted, so that the composition of the liquid may be assumed still to conform to the liquidus. Let us consider, in fig. 3.4, the solidification process under these conditions : at temperature T_1 the liquid of composition C will first deposit crystals of composition c_1 as before. As the

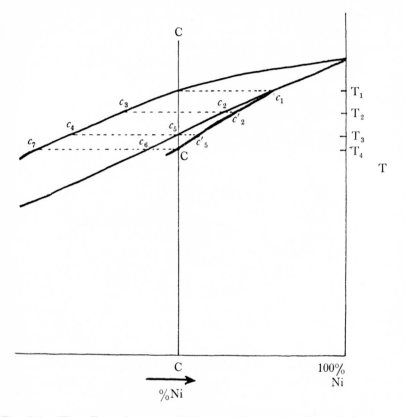

Fig. 3.4. The effect of non-equilibrium cooling on the form of the phase diagram of fig. 3.3.

temperature falls to T_2, the liquid composition will follow the liquidus to c_3, but the layer of solid crystal (composition c_2) deposited at this temperature will not have had time at this fast rate of cooling to inter-diffuse with the nickel-rich material beneath, so that the 'average' composition of the dendrite will be given by c_2', and a concentration gradient will exist in the crystal. Similarly at T_3 the liquid will be of composition c_4, the crystal *surface* of the composition c_5, but the average crystal composition will be c_5' (due again to the inadequate time for diffusion). Solidification will not be complete

46

until T_4, when the last inter-dendritic liquid of composition c_7 is frozen to solid c_6 : this brings the *average* composition of the solid to C, the starting composition.

The locus of the solidus line is thus *depressed* (along c_1, c_2', c_5', etc.) compared with its position under equilibrium conditions (along c_1, c_2, c_5, c_6, etc.), and secondly the structure of the resulting solid is now inhomogeneous and is said to be *cored*. Each crystal will consist of layers of changing composition—the ' arms ' of the original dendrite being richer in the higher-melting consituent (in this case, nickel) than the average, and the inter-dendritic regions being richer in the other constituent (i.e. copper) than the average. In a micro-section

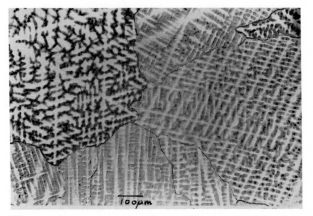

Fig. 3.5. The cored microstructure of a rapidly cooled solid solution of 30% zinc in copper. (Courtesy of the Copper Development Association.)

of this structure therefore, each grain will show a chemical heterogeneity which will be reflected in its rate of attack by the etchant, and fig. 3.5 illustrates this effect in an alloy of copper and zinc. Depression of the solidus and ' coring ' are common features of many cast alloys.

If the cored structure is undesirable, it may be removed by long heat treatments at high temperatures (known as 'homogenization treatments') which allow the solute atoms to be redistributed by solid-state diffusion.

3.3.2. *No mutual solid solubility (simple eutectic)*

The cadmium–bismuth system is a simple eutectic system (see fig. 3.6) which exhibits no solubility of cadmium in bismuth or of bismuth in cadmium. The phase diagram therefore consists of a liquidus line showing a minimum at the eutectic temperature, which is itself marked by a horizontal line. Since the solid phases formed consist simply of pure cadmium or pure bismuth, the *solidus* lines are coincident with the two vertical temperature axes.

Consider first the solidification of an alloy containing 40 wt. % cadmium (alloy (1) in fig. 3.6) : it is liquid at temperatures above 144°C, and on cooling to this temperature it freezes isothermally to give an intimate mixture of cadmium and bismuth metals known as a

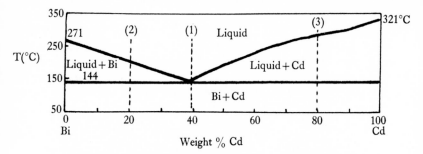

Fig. 3.6. The phase diagram of the bismuth–cadmium system.

'eutectic mixture' with the individual crystals in the form of plates or rods or tiny particles.   Such a structure is sketched in fig. 3.7 *a*. The word 'eutectic' comes from the Greek for 'easily melted', and obviously a mixture of this composition has the lowest melting-point of any cadmium–bismuth mixture.

Considering next alloy (2) in fig. 3.6, which contains 20 wt. % of cadmium, on crossing the liquidus this will start to solidify, when

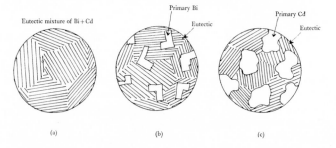

Fig. 3.7. Sketches of a typical microstructure of alloys of composition (*a*) 1, (*b*) 2 and (*c*) 3 in fig. 3.6.

crystals of pure bismuth will separate (the isothermal only intersects the vertical, pure bismuth, solidus), causing the liquid to become enriched in cadmium, and its composition follows the line of the liquidus as the temperature falls.   At 144°C the bismuth crystals will be in equilibrium with liquid which has achieved eutectic composition : the liquid then freezes to form a eutectic mixture of crystals, giving the microstructure (illustrated in fig. 3.7 *b*).   Fig. 3.7 (*c*) illustrates the microstructure of alloy (3) in fig. 3.6.

48

3.3.3. *Limited mutual solid solubility*

(a) *A eutectic system*

There are a number of pairs of metals which show limited mutual solid solubility and which also form a eutectic system. The system lead–tin is probably the most well known, and the phase diagram for this system is shown in fig. 3.8. Here the liquidus *ecf* shows a

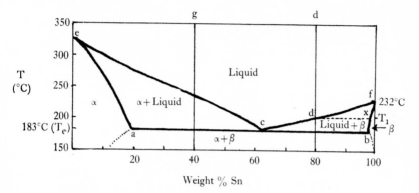

Fig. 3.8. The phase diagram of the lead–tin system.

eutectic minimum at *c*, which means that an alloy containing 38 wt. % lead will remain liquid to a relatively low temperature (183°C), and this is the composition of the familiar lead–tin *solder*.

ea and *fb* are the solidus lines in fig. 3.8; the lead-rich solid solution is labelled the α phase and the tin-rich solid solution is termed the β phase (by convention Greek letters are generally used on phase diagrams to designate the various solid phases). In interpreting the microstructures produced when alloys of various compositions are allowed to solidify, the reasoning will be a combination of those presented above in §§ 3.3.1 and 3.3.2.

Alloys with a tin content between 0 and *a* in fig. 3.8 and between *b* and 100 will simply freeze to the single-phase α and β solid solutions respectively when the temperature falls slowly. Following the reasoning of § 3.3.1, for a given alloy composition, solidification will start when the *liquidus* is crossed and be completed when the appropriate *solidus* is crossed. An alloy of composition *c* will solidify at the eutectic temperature T_e to form a finely divided mixture of the α and β crystals.

However, considering the solidification of alloy *d* (fig. 3.8), at temperature T_1, β crystals of composition *x* will nucleate and as the temperature falls towards T_e, the β crystals grow and change their composition along the solidus *fb* as the liquid phase composition follows the line *dc*. When the temperature reaches T_e, β crystals of composition *b* are in equilibrium with liquid of eutectic composition.

49

This liquid then freezes to an α/β mixture and the microstructure will appear as in fig. 3.7 b or 3.7 c, except that the primary phase will consist of dendrites of a solid solution instead of a pure metal.

Problem 1 (b) on p. 62 relates to the system illustrated in fig. 3.8.

Non-equilibrium conditions

If the liquid alloy is allowed to cool too quickly for equilibrium to be maintained by diffusional processes, one might expect to observe cored dendrites of α or β phase, as discussed in § 3.3.1. The depression of the solidus under these conditions may, however, give rise to a further non-equilibrium microstructural effect if the composition of

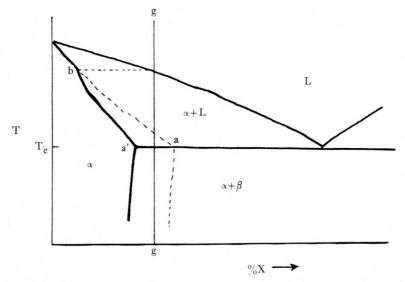

Fig. 3.9. Depression of solidus due to rapid cooling leading to presence of metastable phase.

the alloy is approaching the limit of equilibrium solid solubility (e.g. g in fig. 3.9).

If, due to rapid cooling, the solidus line is depressed from *ba* to *ba'*, alloy g would show some eutectic in its structure, whereas under conditions of slow cooling it would simply freeze to a single phase, as predicted by the equilibrium phase diagram. An experienced metallographer can usually identify this effect, which is quite common in metal castings. In cast tin bronzes, for example, which are essentially copper–tin alloys, particles of hard second phases are often present (which can improve the mechanical properties of the material), even though the equilibrium phase diagram would predict a single-phase copper-rich solid solution for the compositions of the commonly used casting alloys.

50

(b) A peritectic system

Figure 3.10 illustrates a second important way in which two solid solutions may be inter-related on a phase diagram. Temperature T_p is known as the *peritectic* temperature, and the boundaries of the β phase *fd* and *gd* are seen to come to a point at this temperature.

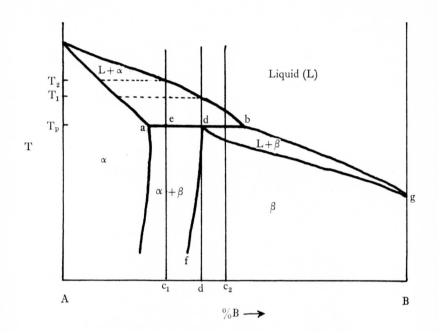

Fig. 3.10. The phase diagram of a peritectic system.

An alloy of composition *d* is said to have the peritectic composition, and we will now examine the nature of the phase change in more detail.

Freezing of alloy *d* will start at temperature T_1 by the separation of crystals of the α solid solution; as the temperature falls under equilibrium conditions, the composition of the solid solution will follow the line of the solidus to *a* and that of the liquid will follow the line of the liquidus to *b*.

At T_p the liquid of composition *b* and the solid α phase of composition *a* react to form the solid solution β phase. Having for simplicity chosen an alloy composition corresponding to the peritectic point, the reactants are fully consumed so that β is the only structure observed below the temperature T_p.

Considering now an alloy of composition c_1, this will likewise freeze initially at temperature T_2 to form α in the liquid phase, but at

51

temperature T_p, although the reactants for the peritectic reaction are present (i.e. α phase of composition a and liquid of composition b), by the application of the lever rule it is seen that the fraction of solid phase present eb/ab is greater than that required for the peritectic reaction to proceed to completion, so that the β phase will be produced (with the disappearance of all the liquid phase and part of the α phase), and the microstructure will consist of ' walls ' of the peritectically produced β phase enveloping the unconsumed parts of the original α dendrites.

Problem 1 (*c*) on p. 62 relates to the system illustrated in fig. 3.10.

Non-equilibrium conditions

When the peritectic reaction begins, the α phase and the liquid are in contact and the β phase is formed at the solid–liquid interface, as illustrated in fig. 3.11. When the β phase has formed an envelope

Fig. 3.11. The microstructure of a specimen showing a peritectic reaction : the pale grey phase (X) has reacted with the liquid (Y) to produce the white phase (Z) in an aluminium–copper alloy.

about the α phase, the rate of reaction will depend upon the rate of diffusion of the reactants through the wall of β phase that separates them, and since this may be a sluggish process it is quite commonly observed in cast alloys which have not been cooled extremely slowly that the peritectic reaction has not gone to completion, and other ' *metastable* ' phases are seen.

3.3.4. *Phase transformations in the solid state*

Transformations in the solid state are much more subject to the rate of atomic migration as well as to the phase diagram than are liquid–solid transformations, and one important feature of this type of reaction is the possibility of being able to suppress it by rapid cooling and then to re-introduce it at a later time by re-heating the material.

(*a*) *Changing solid solubility with temperature*

Figure 3.12 illustrates this feature in part of a eutectic diagram for a mixture of two metals. Considering an alloy of composition *c*, this will solidify to a single-phase solid solution, α, which is stable only

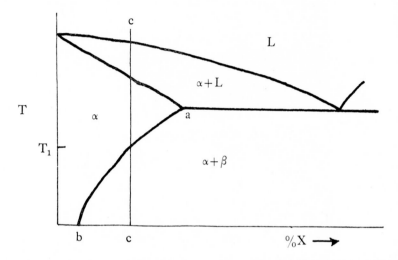

Fig. 3.12. Phase diagram showing decreasing solid solubility with decreasing temperature.

down to temperature T_1. The line *ab* is called a 'solid solubility' or 'solvus' line, and in the present example it shows that the solubility of element X in the α phase falls from a value of *a*% at the eutectic temperature to *b*% at the lowest temperature on the ordinate axis. Many commercially important alloys feature this decreasing solid solubility with temperature, for example, the high strength aluminium alloys used in the construction of airframes, and the reasons for this will emerge in Chapter 5.

As the temperature falls below T_1 (fig. 3.12), the α phase crystals contain more of element X than they would do at equilibrium— that is they become *supersaturated*. If the cooling is slow, crystals of the β phase then form. In a supersaturated liquid solution the

53

crystals can form anywhere, but the solid medium puts severe constraints on their location. For example, it might be expected that the initial precipitation of the β phase would take place along the grain boundaries of the original α phase—firstly because the atoms are more loosely held in the grain boundaries and so might be expected to 'break away' more readily to form the new phase, and secondly because the atomic disarray at the α-phase grain boundaries could help to accommodate any local volume changes associated with the growth of the new β crystals. Figure 3.13 illustrates this type of solid-state

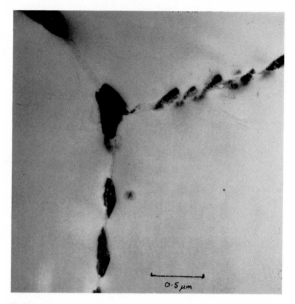

Fig. 3.13. Solid-state precipitation in a two-phase alloy, showing grain-boundary precipitate. Electron micrograph.

precipitation in stainless steel, and particles of the second phase are seen to have formed along the grain boundaries of the primary phase.

In many alloy systems the change in solubility with temperature is so great that the second (β) phase cannot all be accommodated in the grain boundaries of the primary (α) phase, and precipitation *within* the primary grains then occurs. This 'intragranular' precipitation is usually found to take the form of plates or needles in parallel array (see fig. 1.1). This striking geometrical feature arises from the tendency of the new crystals to grow with their interfaces aligned parallel with certain definite crystal planes of the primary (α) phase. These planes will be such that there is a better atomic fit across the α/β interfaces than if the β phase was randomly distributed inside the α phase. Because of the high symmetry of most metal crystals there

will usually be several equivalent sets of ' planes of good matching '
along which the second phase can form, leading to the presence of
several ' families ' of parallel platelets of the second phase dispersed
within the parent crystals.

Let us now return to consider the micrograph of the meteorite
shown in fig. 1.1 : at high temperature it consisted of a single crystal
of solid solution, then as it cooled after reaching the Earth, solid-state
precipitation of iron-rich crystals took place, and the striking geometric
array developed, with the new crystals lying parallel to certain sets of
crystal planes of the parent crystal. Alois de Widmanstätten first
observed this effect in a meteorite which had fallen in 1751 near
Zagreb in Jugoslavia, and today any solid-state precipitate which
shows this characteristic parallel array is known as a ' Widmanstätten
structure '.

(b) Eutectoid and peritectoid processes

We saw in §§ 3.3.2 and 3.3.3 that eutectic and peritectic phase
changes may proceed as follows :

$$\text{eutectic :} \qquad \text{liquid phase} \underset{\text{heating}}{\overset{\text{cooling}}{\rightleftarrows}} \text{solid A} + \text{solid B} ;$$

$$\text{peritectic :} \qquad \text{liquid phase} + \text{solid A} \underset{\text{heating}}{\overset{\text{cooling}}{\rightleftarrows}} \text{solid B}.$$

Processes analogous to these can take place wholly in the solid state,
and are then called ' eutectoid ' and ' peritectoid ' respectively :

$$\text{eutectoid :} \qquad \text{solid A} \underset{\text{heating}}{\overset{\text{cooling}}{\rightleftarrows}} \text{solid B} + \text{solid C} ;$$

$$\text{peritectoid :} \qquad \text{solid A} + \text{solid B} \underset{\text{heating}}{\overset{\text{cooling}}{\rightleftarrows}} \text{solid C}.$$

No new principles are involved in these changes, and we will not
discuss them in detail here, although it must be emphasized that,
being wholly in the solid state, and dependent upon atomic migration
to proceed, they may well be suppressed if the alloy is cooled quickly.

3.4. Control of phase distribution in alloys

Thermal treatment can be used to control the size and distribution
of second-phase particles in any alloy which undergoes a phase
transformation in the solid state.

Alloy systems which have a phase diagram showing a decreasing
solid solubility limit with decreasing temperature (e.g. fig. 3.12) are
particularly appropriate for these treatments, and particles of second
phase can often be made to precipitate in a very finely dispersed form.
In fig. 3.12 an alloy c exists as a single-phase α solid solution at high

temperatures (above T_1) but on slow cooling below T_1 it becomes supersaturated with respect to the second (β) phase, which therefore separates out.

The distribution of the β phase may be controlled in the following general way.

The alloy is first solution heat-treated at the high temperature (above T_1) and then rapidly cooled by quenching into water or other cooling fluid. Solid-state diffusion is suppressed in this way, so that

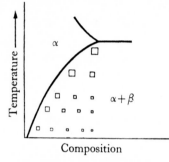

Fig. 3.14. Variation of precipitate size with ageing temperature.

the β phase cannot separate and the alloy exists at the low temperature in an unstable supersaturated state. If the temperature is now increased, so that diffusion can take place at a measurable rate, the second phase will nucleate and grow.

In alloys of relatively low melting-point (in aluminium alloys, for example) there will be an appreciable diffusion rate of solute atoms at room temperature, so that over a sufficient length of time, the second phase will precipitate out and form a fine Widmanstätten pattern. This effect is known as ' ageing ', but in most alloys the temperature has to be raised in order to cause precipitation to occur and the material is said to be ' artificially aged '. The rate of growth of the precipitate is controlled by the rate of atomic diffusion, so that the precipitation increases with increasing ageing temperature, but the size of the precipitate becomes finer as the ageing temperature is lowered, as shown schematically in fig. 3.14.

Quenching and ageing are therefore very powerful means of controlling the distribution of a precipitate of second phase in an alloy. After quenching a solution-treated alloy, a high ageing temperature is selected if a coarse, widely spaced dispersion of particles is required, and a lower ageing temperature is used to produce the second phase in a more finely divided form.

If, at any temperature, the *time* of heat treatment is very prolonged, coagulation or coarsening of the particles occurs : the small ones tend to redissolve and the large ones to grow at their expense. Thus the many finely dispersed small particles are replaced by fewer, coarser particles which will be more widely spaced apart, and the alloy is said to be in the *over-aged* condition. Problem 2 (*a*) on p. 62 is concerned with this situation.

3.5. *The structure of some simple steels*

Because of the great importance of steel as a structural material, we will conclude this chapter with an outline of the metallography of some simple steels which will also provide some further useful illustrations of the application of the general principles of phase diagrams set out in the preceding sections.

Steel is basically an alloy of iron and carbon, and one of the most important properties of this material is its ability to have its structure and hence its properties changed by heat treatment. A detailed discussion of the solid-state phase transformations in steels is beyond the scope of this text, so that only an introductory outline will be given.

The iron-rich end of the iron–carbon phase diagram is given in fig. 3.15, and although it may at first glance seem much more complicated than anything we have considered so far, provided a systematic approach is made, it can be treated as simply as the previously considered diagrams. The liquidus forms a simple eutectic at 4·3 wt. % carbon, and in the solid state iron exists in three crystal structures : δ (b.c.c.), stable between 1535°C and 1400°C, γ (f.c.c.), stable between 1400°C and 910°C and α (b.c.c.), stable below 910°C. Each of these three structures can be seen to have different solubilities for carbon : only the γ (f.c.c.) phase shows appreciable solubility (up to approximately 2 wt. % carbon at 1150°C), forming a solid solution known as ' austenite '. The body-centred solid solutions are known as α and δ ' ferrite '.

The solid phase in equilibrium with the iron-rich solid solutions has the composition Fe_3C and is known as ' cementite ', and the phase diagram as a whole may be considered as three interlinked simple ones : the top left-hand corner being a simple peritectic (cf. fig. 3.10), the liquidus a eutectic (cf. fig. 3.8) between γ (austenite) and Fe_3C (cementite), and the lines GSE forming a eutectoid—describing the decompositon of the γ solid solution at 723°C to a mixture of α (ferrite) and Fe_3C (cementite).

We will use the phase diagram of fig. 3.15 to explain the microstructures of ' plain carbon ' steels (i.e. steels free from further special alloying additions). These steels are conveniently divided into three groups : low-carbon or mild steels containing less than 0·3 wt. % C, medium-carbon steels containing between 0·3 to 0·7 wt. % C, and high-carbon steels containing between 0·7 to 1·7 wt. % C. The low-carbon steels have moderate strengths but are easily fabricated and are used in very large amounts (i.e. about 90% of all steel is in this category) for structural purposes. The medium and high-carbon steels are heat-treated after fabrication in order to develop the high strength and toughness required in engineering components.

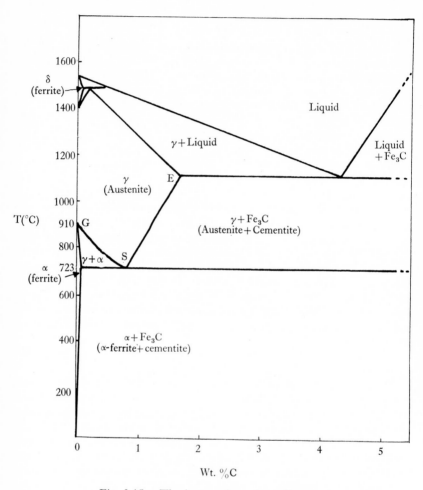

Fig. 3.15. The iron–carbon phase diagram.

3.5.1. *Slowly cooled structures*

The most important reaction in steel is the decomposition of the austenite (γ phase) on cooling. Consider the slow cooling of a steel of 0·8 wt. % C content (i.e. of the eutectoid composition); at 723°C the structure will transform to a eutectoid mixture consisting of alternate lamellae or plates of α ferrite and cementite. This microconstituent is known as 'pearlite', and its appearance is sketched in fig. 3.16 a. A steel of higher carbon content (known as a 'hyper-eutectoid' steel), say of 1 wt. % C, will remain austenitic down to the temperature around, say, 800°C, at which the solvus line SE in fig. 3.15 is crossed, so that Fe_3C will be precipitated at the austenite grain boundaries (fig. 3.16 b). When the temperature falls to below 723°C, the residual austenite will transform to pearlite, and the microstructure will appear as illustrated in fig. 3.16 b.

A low-carbon steel (i.e. a 'hypoeutectoid' steel) will transform when the temperature falls below the line GS in fig. 3.15 by the

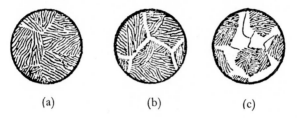

(a)   (b)   (c)

Fig. 3.16. (a) The eutectoid structure of α iron and Fe_3C known as 'pearlite'. (b) The structure of a high-carbon steel, showing precipitation of Fe_3C in the original γ-Fe boundaries, enclosing regions of pearlite. (c) Low-carbon steel structure : area of α-Fe enclosing regions of pearlite.

precipitation of α ferrite at the austenite grain boundaries (fig. 3.16 c) and once more at 723°C the remaing austenite will transform to pearlite.

Problem 2 (b) on p. 63 is concerned with transformation in a slowly cooled steel.

3.5.2. Quenched structures

The above microstructures form in plain carbon steels which have been moderately slowly cooled (e.g. by cooling in air) from temperatures within the austenitic phase field, say from 50°C above the line GSE in fig. 3.15. This is called a 'normalizing' heat treatment ; medium and high-carbon steels are very commonly subjected to more complex treatments in order fully to exploit their properties. These treatments involve, firstly, heating the alloy into the austenite phase field as before but then *quenching* it into water or brine which suppresses diffusion and thus the formation of ferrite and cementite. Under these conditions the austenite transforms by a process *not involving diffusion* into a metastable distorted form of b.c.c. iron known as 'martensite'. The phase change takes place by the *shearing* of lens-shaped regions of each austenite grain to form martensite, as shown in the sketch fig. 3.17 a. This process is extremely rapid, and the transformation may be completed in a few microseconds, and takes place in a way very similar to 'mechanical twinning' to be discussed on p. 72.

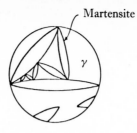

(a)

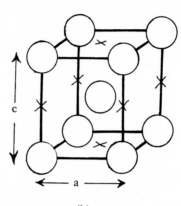

(b)

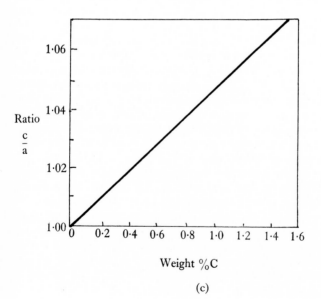

Weight %C

(c)

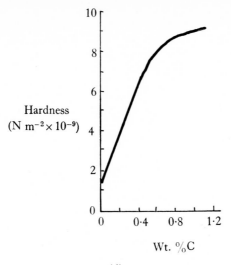

Hardness
(N m^{-2} × 10^{-9})

Wt. %C

(d)

Fig. 3.17. (*a*) The microstructure of martensite (partly transformed from austenite). (*b*) The crystal structure of martensite : a body-centred tetragonal unit cell with interstitial carbon occupying a proportion of the sites marked ×. (*c*) Showing increasing tetragonality of the martensite structure with increasing carbon content. (*d*) showing linear increase in hardness of martensite with increase in carbon content.

All the carbon originally dissolved in the austenite at high temperature remains after quenching in interstitial solution in the martensite crystals, occupying the positions marked × in fig. 3.17 *b*. This has the effect of distorting the lattice from cubic to *tetragonal* symmetry ; one of the cube axes becomes slightly longer and the other two equal ones slightly shorter. The degree of tetragonality increases as ratio c/a increases (fig. 3.17 *b*), and the effect of increasing carbon content is to increase the value of c/a as an increasing proportion of the × sites are occupied. This is shown graphically in fig. 3.17 *c*; in the absence of carbon, c/a = 1 and the crystal is cubic (i.e. ferrite, or body-centred iron), and increasing tetragonal distortion is observed as the carbon content rises.

This local lattice distortion due to the dissolved carbon has the effect of hardening the structure, and fig. 3.17 *d* illustrates the effect. A steel quenched to form martensite will therefore be hard (due to its state of internal strain), but it will also be *brittle*, so we cannot usually exploit the increased strength. A second heat treatment, called *tempering*, is required to change the structure to one with lower hardness but less brittleness.

Martensitic transformations are not limited to the iron–carbon system, but are found in many other systems such as Cu–Al, Fe–Ni and Au–Cd. In contrast to other phase transformations which proceed by the nucleation and growth of new crystals by diffusion, martensitic transformations are rapid, diffusionless and involve a shearing of lattice planes to form the new arrangement.

3.5.3. *Tempered martensite*

If a martensitic steel is reheated within the temperature range 200–600°C (i.e. below temperatures at which austenite can re-form) it rapidly decomposes to form b.c.c. ferrite and particles of cementite (Fe_3C). The dissolved carbon is thus rejected as a carbide precipitate so that the lattice is restored to cubic symmetry once more, as expected from fig. 3.17 c. The two-phase microstructure which develops after tempering is on an extremely fine scale : the size of the carbide particles produced depends upon the tempering temperature and time of treatment. In practice tempering times are of the order of 30 minutes, and the carbide particle size is of the order 1 μm after high (500–600°C) tempering temperatures (and therefore just resolvable in the optical microscope), but may be less than 10^{-2} μm after tempering at low temperatures (200–300°C). Until the electron microscope was applied to metallic structures, it was not realized that a change in tempering temperature produces essentially only a change in scale of the structure of the tempered martensite and not in its fundamental character. The result of this was that the early metallographers were misled into giving special names (e.g. ' secondary troostite ') to some of these structures (which were in fact unresolvable in their instruments), and the wide use in early textbooks of these illogical names led to the confusion of subsequent generations of students.

3.6. *Problems and exercises*

3.6.1. *Liquid–solid transformations*

(*a*) In fig. 3.3, consider the changes that occur when an alloy of composition C is heated from room temperature to 1400°C.

(*b*) In fig. 3.8, consider the changes that occur when an alloy of composition g is allowed to solidify. By applying the lever rule, estimate the relative proportions of primary phase and of eutectic that you would expect to observe metallographically.

(*c*) In fig. 3.10, by applying the lever rule, estimate the fractions of α phase and of liquid phase required to react completely to form β. Consider the changes that occur when an alloy of composition c_2 is allowed to solidify.

3.6.2. *Solid-state transformations*

(*a*) Sketch the microstructure of a precipitated two-phase alloy (e.g. composition c in fig. 3.12) (i) before and (ii) after an over-ageing treatment.

What is the driving-force for the over-ageing process?

(b) Consider a solid solution of γ (austenite) in fig. 3.15. As the temperature is slowly decreased, discuss how the composition of the austenite changes in the temperature range between the first appearance of the grain-boundary phase (i.e. ferrite or cementite, depending upon the composition of the steel) and the transformation to pearlite.

Supplementary reading list

Introduction to the Science of Metals: M. H. Richman (Ginn Blaisdell), Chapters 7–9.

Modern Physical Metallurgy: R. E. Smallman (Butterworths), Chapter 2.

Metals Reference Book: C. J. Smithells (Butterworths) is a useful source of equilibrium diagrams in metallic systems.

CHAPTER 4

deformation

4.1. *Introduction*

THE mechanical properties of a metal are closely related to its microstructure—for example, the strength increases as the grain size is reduced, and in two-phase alloys the strength increases as the spacing between the particles of the second phase is reduced. A main branch of study in physical metallurgy is that of the principles relating the structure and the mechanical properties, so that, ideally, by understanding these principles, materials may be produced with any desired properties.

Having discussed the crystal structures encountered in metals and alloys, we will now examine the ways in which they can undergo deformation, and it is important at the outset to distinguish between *elastic* deformation (after which, when the deforming force is removed, the solid spontaneously regains its original size and shape), and permanent or *plastic* deformation (which is not restored in this way). For example, an ordinary metal paper clip can be slightly deformed so that it springs back elastically to its original shape, but if it is bent sufficiently it undergoes a permanent change of shape, by plastic deformation.

4.2. *Stresses and strains*

In order to allow for the physical dimensions of the body to which they refer, stresses and strains, rather than forces and extensions, are used to describe the deformation behaviour of a solid.

4.2.1. *Stresses*

Consider a rod of cross-sectional area A ; if a force F parallel to the axis of the rod is applied to the ends of the rod, the force per unit area of cross-section is called the *normal stress* acting, and is denoted by the symbol σ, where $\sigma = F/A$. In addition, this stress would be termed ' tensile ' or ' compressive ' according to whether it tended to extend or to shorten the rod.

Consider now a rectangular block of material (fig. 4.1 *a*) with a horizontal force F applied parallel to its upper surface whose area is A. The force per unit area on this surface is called the *shear stress* and is denoted by the symbol τ, where $\tau = F/A$.

64

4.2.2. *Strains*

Just as it is convenient to express force in terms of stress (force per unit area) so is it convenient to express deformation with reference to the dimensions of the body.

If a rod of length l_0 (fig. 4.1 *b*) extends (or contracts) to a length l when acted upon by a stress, the *linear strain* is defined as the

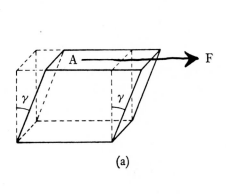

(a)

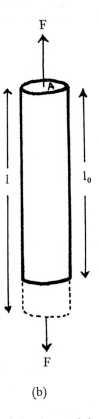

(b)

Fig. 4.1. (*a*) Shear stress (F/A) and shear strain (γ). (*b*) Linear stress (F/A) and linear strain $(l-l_0)/l_0$.

elongation (or contraction) per unit length and is denoted by the symbol ε, where $\varepsilon = (l-l_0)/l_0$.

The deformation resulting from a shear stress involves a shape change ; the block of material shown in fig. 4.1 *a* is deformed so that the front and rear faces become parallelograms, and the *shear strain* is the angle γ expressed in radians. Both types of strain are therefore seen to be expressed as a dimensionless parameter.

4.3. *Elastic deformation*

The elastic behaviour of crystalline substances can be deduced by considering a pair of atoms which have an equilibrium separation a_0.

The potential energy (ϕ) of these atoms will vary as their distance of separation (r) changes, and this may be written in the form :

$$\phi = -\frac{A}{r^a} + \frac{B}{r^b}. \qquad . \qquad . \qquad . \quad (4.1)$$

This expression is illustrated graphically in fig. 4.2 a, and these are known as the Condon–Morse curves. The value of r corresponding to the minimum potential energy is the equilibrium spacing a_0,

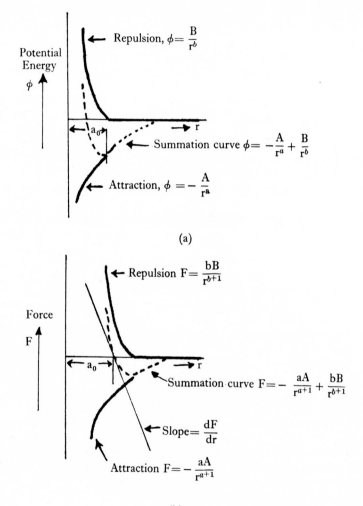

(a)

(b)

Fig. 4.2. Condon–Morse curves showing the qualitative variation of (a) energy and (b) force with distance of separation r of atoms.

66

determined by the relative magnitudes of the attractive interaction (given by $\phi = -A/r^a$) and the repulsive interaction (given by $\phi = +B/r^b$). The *forces* of attraction and repulsion (F) may be derived from equation (4.1), since

$$F = -\frac{\partial \phi}{\partial r} = -\frac{aA}{r^{a+1}} + \frac{bB}{r^{b+1}}. \qquad . \qquad . \qquad . \qquad (4.2)$$

Comparison of equations (4.1) and (4.2) indicates that the curves must have the same form, and this is shown in fig. 4.2 *b*. The net force is zero at a_0, and displacement in either direction will call restoring forces into play. The normal range of elastic displacements rarely exceeds $\pm 0.5\%$, and fig. 4.2 *b* shows that the tangent $\partial F/\partial r$ very nearly coincides with the force curve for such small departures from a_0. For all practical purposes, therefore, it may be stated that *the stress is a linear function of the strain.*

Although this discussion has been concerned with an isolated pair of atoms, the same type of behaviour is exhibited by the atoms of a crystalline array, and our conclusion expresses the observation known as *Hooke's law.* Hooke observed that at low stresses, for wires stressed in tension, that the stress was proportional to the strain, and an *elastic modulus* can be defined as the ratio stress/strain. The elastic behaviour of crystalline materials in compression is the same as that in tension, as shown by the tangent in fig. 4.2 *b*.

Under these very small elastic strains, *every atom in the crystal* is slightly displaced from its equilibrium position, but a high stress is usually required because the displacement is opposed by the restoring forces, due to the change in length of the primary bonds of the crystal, be they ionic, covalent or metallic in character. Most crystals are *anisotropic* in their elastic behaviour, with different values of elastic modulus being obtained when strains are applied in different crystallographic directions. In polycrystalline metals, however, the individual anisotropy of the randomly oriented grains is cancelled out in the bulk, and an average value of the elastic constants is obtained. Engineers are therefore justified in assuming that such materials are elastically isotropic.

4.3.2. *The elastic moduli*

Three important elastic proportionality constants are in common use, which differ only in the types of stress and strain which they relate.

Young's modulus $(E) = \sigma/\varepsilon$. Most school physics laboratories are equipped to make an experimental determination of E, using a specimen in the form of a long wire.

(Question : Why is a *long* specimen chosen ?)

Shear modulus (μ or G) $= \tau/\gamma$. A *torsion* experiment can be used to measure G, whereby a cylindrical specimen is twisted about its longitudinal axis under a known applied couple.

Bulk modulus $(K) = \sigma_{hyd}/(\Delta V/V_0)$. Here σ_{hyd} is the hydrostatic tensile or compressive stress (which would be produced, for example, if the solid were immersed in a liquid which was enclosed in a cylinder by a piston under high external pressure) and $\Delta V/V_0$ is the fractional volume expansion or contraction.

Although a knowledge of the numerical value of these constants is important to the engineer, in structures where lightness is important, such as in aircraft, the 'specific moduli' are often more important than the actual moduli. The former are obtained by dividing the moduli by the specific gravity, and some of the materials of highest specific Young's modulus are :

$$\left.\begin{array}{ll}
\text{diamond} & 24 \times 10^{10} \\
\text{beryllium} & 16 \times 10^{10} \\
\text{silicon carbide} & 13 \times 10^{10} \\
\text{alumina} & 9 \times 10^{10} \\
\text{cellulose fibre} & 6 \cdot 9 \times 10^{10}
\end{array}\right\} \text{N m}^{-2}.$$

The following materials are of somewhat lower specific modulus :

$$\left.\begin{array}{ll}
\text{glass} & 2 \cdot 76 \times 10^{10} \\
\text{steel} & 2 \cdot 62 \times 10^{10} \\
\text{aluminium and wood} & 2 \cdot 55 \times 10^{10} \\
\text{magnesium} & 2 \cdot 48 \times 10^{10}
\end{array}\right\} \text{N m}^{-2}.$$

4.3.3. *Factors affecting the elastic constants*

The elastic constants are among the most structure-insensitive of the mechanical properties, because they depend essentially on the magnitude of the interatomic binding energy. The elastic moduli are reduced by an increase in temperature, and if the crystal structure of the material is changed when the temperature is raised (due to an allotropic change in a pure metal) then the moduli can change discontinuously at the temperature of the phase change.

Alloying alters the energy of the atomic bonds and can lead to an increase or a decrease in modulus with respect to the solvent. In metals which are completely soluble in each other in the solid state, the moduli vary linearly with change in composition across the phase diagram. In systems which form intermediate phases, considerable positive or negative deviations from a mixture law are, however, observed.

4.4. *Plastic deformation*

4.4.1. *The limit of proportionality*

When a test-piece of polycrystalline metal, such as copper or aluminium, is subjected to an increasing tensile stress σ, its initial response is as shown in fig. 4.3. At first the material extends elastically

(AB), and Young's modulus is given by the slope of this part of the curve of stress against strain. At stresses above σ_L, which is the *limit of proportionality* of the material, the deformation is no longer elastic but plastic, and slope of the stress–strain curve deviates from the value of E.

As the sample elongates plastically, the stress required continues to increase, and the material is said to *work-harden* or *strain-harden*. If

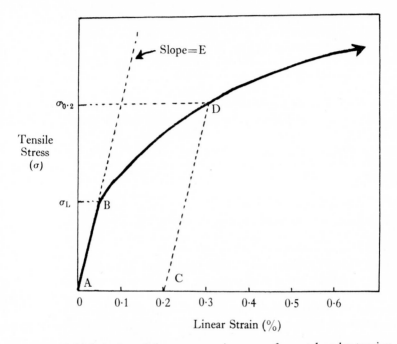

Fig. 4.3. Initial portion of the stress–strain curve of a metal under tension.

the specimen were unloaded at any point during the test it would contract elastically but, when unloaded, show a permanent extension. For example, if the load were removed when point D (fig. 4.3) was reached, the specimen would contract along the dotted line DC (which is parallel to BA) and be permanently extended by a strain of 0·2%. On re-straining the material, it would stretch elastically along CD until at D it would begin to flow plastically once more and to follow the stress–strain curve. The term *yield stress* is sometimes used to describe the stress (σ_L) at which elastic flow first begins, and *flow stress* is used to describe the stress at which a strained specimen will flow plastically (e.g. at the point D in fig. 4.3 after a strain of 0·2%).

As it is experimentally difficult to determine a precise value for σ_L (assessing the point of departure from linearity on the curve demands highly sensitive strain measurement), engineers usually quote an arbitrary stress which will produce a known, small amount of plastic flow. This stress is known as the ' proof stress ' or ' offset yield stress ', and the most common of these is that we have chosen in the illustration of fig. 4.3, the stress at 0.2% plastic strain, $\sigma_{0.2}$. This can readily be obtained from any stress–strain curve by constructing a line (CD) parallel to the elastic part of the curve (AB) from the point on the abscissa corresponding to a strain of 0.2%.

4.4.2. *The mechanisms of plastic flow in crystals*

Most pure metals are characterized by their high *ductility*, which enables them to be easily fabricated into their required shape by forging processes. This property also, however, confers *toughness*, which protects metals from catastrophic brittle failure when they are deformed beyond their elastic limit, in contrast to the behaviour of useful brittle materials such as ceramics and glasses, which are strong and *fragile*.

By studying the deformation behaviour of single crystals the mechanisms of plastic flow in metals are generally well understood, and the two most important processes are known as *slip* and *twinning*.

(a) *Deformation by slip*

By controlled solidification of the molten metal it is not difficult to prepare a single crystal of a hexagonal metal such as cadmium in the form of a wire of a few millimetres' diameter, and several centimetres in length. If it is extended plastically, it will appear as shown in fig. 4.4 *a*, and fig. 4.4 *b* illustrates diagrammatically the process which has taken place. This type of plastic flow is called ' slip ' since it occurs by the slipping of crystal planes over each other. Whatever the direction of the applied force, the direction of crystal slip is found to be always limited to certain crystallographic ' slip directions '. The slip direction selected is one in which the interatomic spacing is a minimum, such as the close-packed directions illustrated in fig. 1.6. In many metals, such as the cadmium crystal illustrated, the crystal *planes* upon which slip occurs are also crystallographic and are termed ' slip planes '. In metals of h.c.p. and f.c.c. crystal structure, for example, slip nearly always takes place on the close-packed planes (see figs. 1.7 and 1.8).

An ordinary piece of polycrystalline metal will also deform by the process of slip within each individual grain, although the grain boundaries are regions where slip will not readily occur, so that a polycrystal will be harder to deform than a single crystal. It is simple to demonstrate the process by preparing a sample as for normal metallographic examination and then subjecting it to slight plastic deformation ; fig. 4.5 *a* is a typical example of the resultant structure in the case of aluminium. Slip is seen to have occurred by the presence of ledges on the surface arising from the slip process illustrated in fig. 4.4 *b*, and they are usually called 'slip bands'. Slip is clearly an inhomogeneous process, being confined to the regions of the slip bands, unlike elastic deformation, which is homogeneous throughout the whole crystal.

In many metals a slightly more complicated slip process takes place by slip occurring simultaneously on more than one set of planes.

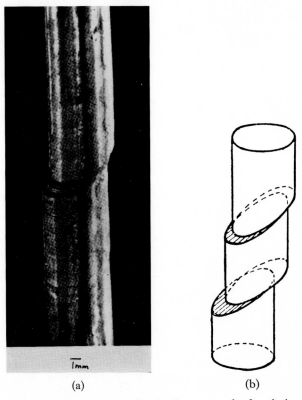

(a)   (b)

Fig. 4.4. Illustration of crystal slip (*a*) in a crystal of cadmium, (*b*) in a diagrammatic form.

Figure 4.5 *b* shows the slip-band distribution after this 'multiple slip' has taken place : in this case two sets of planes have operated as slip planes. Metals with an f.c.c. crystal structure are likely to show multiple slip since each crystal contains four identical families of close-packed planes which, if the applied deformation is in the correct direction, can operate as slip planes.

Experiment 4.6.1 (p. 82) describes how slip bands may be observed on the surface of deformed lead.

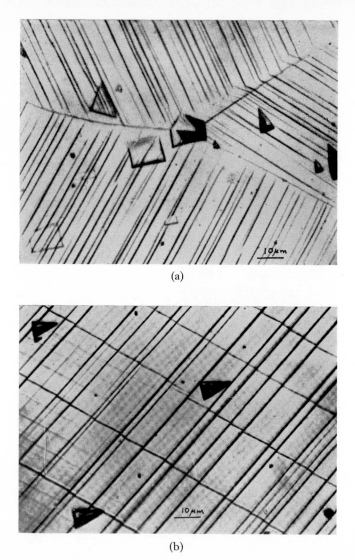

(a)

(b)

Fig. 4.5. Slip bands in aluminium. (*a*) Single slip in each grain. (*b*) Multiple slip. The polygonal surface pits were formed during etching.

(*b*) *Deformation twinning*

Twinning is a much less common feature than slip in plastically deformed metals: an example is shown in fig. 4.6 *a*. The atomic displacements involved are illustrated in fig. 4.6 *b*, and these give rise to bands of crystal of twinned orientation (often lens-shaped) within the grain. The crystallographic nature of deformation twins is the

72

(a)

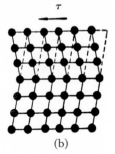

(b)

Fig. 4.6. (a) Mechanical twins in deformed polycrystalline uranium. (b) Diagram showing the atomic movements in twinning, and development of surface tilts.

73

same as that of annealing twins (fig. 2.10), although their origin is quite distinct.

The stress required to cause twinning is usually higher than that for slip, which is why it is less common. Iron can be made to twin by straining very rapidly at room temperature, or more slowly below about 100°K. Copper and other f.c.c. metals can be twinned at extremely low temperatures, but the more striking examples are provided by non-cubic metals such as zinc, tin and bismuth. A creaking sound is heard if a rod of polycrystalline tin is bent plastically (this is called 'the cry of tin'), and it is caused by the bursting into existence of deformation twins. This abrupt formation of twins leads also to *jerks* on the stress–strain curve, so that it appears serrated rather than smooth as shown in fig. 4.3.

Experiment 4.6.3 (p. 83) describes how deformation twinning may be demonstrated with tin.

4.4.3. *Plastic flow processes at the atomic level*

Figure 4.7 illustrates the process of slip at the atomic level. After slip has taken place the original structure is re-formed perfectly, and the original properties of the crystal will be perfectly restored.

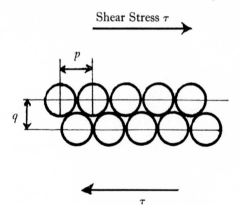

Fig. 4.7. The atomic movements in slip.

Clearly, when the applied force is removed, the crystal will remain stable in the deformed state. Although the slip movement appears to resemble that of a pack of cards which has been pushed lengthwise, the actual mechanism by which slip takes place does not correspond to the sliding of cards. It is found that crystals slip under load at stresses very much below those which would be required to move two perfect crystal planes past one another. This latter stress can be estimated as follows.

Let the spacing between atom centres in the direction of slip be p and the spacing of the layers be q as in fig. 4.7. When the shear stress is τ, let the shear displacement of the upper layer over the lower be x. When the upper layer is in a stable equilibrium position (i.e. $x = 0, p, 2p$, etc.), the shear stress is obviously zero, and it will also be so when the upper layer is displaced by $\frac{1}{2}p$, $\frac{3}{2}p$, etc. to a position of unstable equilibrium, when a slight movement either way in the absence of a shear stress would cause the upper layer to move to one of the stable positions. From one of these stable equilibrium positions, a small displacement will permit Hooke's law to be obeyed, and the shear stress is given by:

$$\tau = G.\gamma.$$

The shear strain γ will be given by x/q.

The curve of τ against x will change sign for every increase in x by $\frac{1}{2}p$, and its shape will depend upon the nature of the interatomic

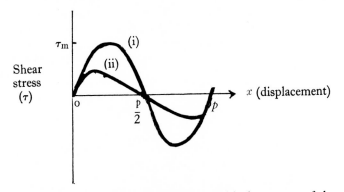

Fig. 4.8. The variation of shear stress (τ) with the amount of shear.

forces. If we assume, as shown in fig. 4.8 (i), that the curve is a sinusoidal function, of period p, we can write:

$$\tau = \tau_m \sin \frac{2\pi x}{p},$$

where τ_m is the maximum value of τ which would have to be applied before slip would occur, i.e. is the shear *yield stress*.

For small displacements, $\tau = \tau_m(2\pi x/p)$, since, for small angles $\sin \theta \approx \theta$, and from Hooke's law : $\tau = G(x/q)$, so that

$$\tau_m = \frac{G}{2\pi} \cdot \frac{p}{q}.$$

Since $p \doteqdot q$, $\tau_m \doteqdot G/2\pi$, which is of the order $1\cdot5 \times 10^{10}$ N m^{-2} for most materials. The observed values of the yield shear stress is of the order 10^5–10^6 N m^{-2}, so there is a discrepancy of 10^4–10^5 between the theoretical and the observed values. In fact, a more likely form of the shear stress–displacement curve is curve (ii) of fig. 4.8. The maximum shear stress for this curve is found to be $\approx G/30$, so that the major part of the discrepancy still exists, even with a refined model.

A more realistic picture of the slip process would be to visualize slip beginning in one small area of the plane and then spreading over the rest of the plane. While this is taking place, the slip plane will be divided into a slipped area and an unslipped area, and a line of demarcation—called a *dislocation line*—will be moving across the slip plane.

The passage of a dislocation line across a slip plane is like the movement of a ruck in a carpet—it is obviously more difficult to make one carpet move bodily over another by pulling one end of it, than to make a ruck in the carpet and to move the ruck along. The ruck separates the ' slipped ' from the ' unslipped ' part of the carpet in the same way that a dislocation line separates the slipped and unslipped regions of a slip plane in a crystal. Calculations show that the stresses required to make dislocation lines move are in good agreement with their measured yield stresses, so that the presence of such faults can account for the weakness of metals. This means that metal crystals are

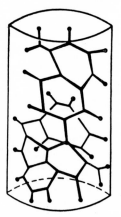

Fig. 4.9. A network of dislocations within a crystal.

imperfect : they contain dislocations before any plastic deformation, due to irregularities in the crystallization process, and the average distance between these grown-in dislocations is a few thousand atom spacings.

The atoms which lie along a dislocation line are in positions of higher energy than those atoms in a region where the crystal array is perfect. This energy causes the dislocations to have an effective ' line tension ' (in the same way that the interface between the phases, or between two crystals of differing orientation, will possess an effective 'surface tension'), so that in an unstressed crystal they will tend to minimize their length and form a network of straight lines within each grain as illustrated in fig. 4.9. A dislocation line cannot end within a crystal—it must either run to a free surface, to a grain boundary or to another dislocation. Some of these dislocations lie in planes that are not favourable for slip, so that they cannot easily be moved when the crystal is stressed, and only those parts of the dislocation network which happen to lie in a slip plane are the ones which move when the crystal is deformed plastically.

Let us first consider (fig. 4.10 a) a small block of crystal through which a simple form of dislocation passes EE, which lies on its slip plane ABCD. When the upper part of the crystal is stressed as shown,

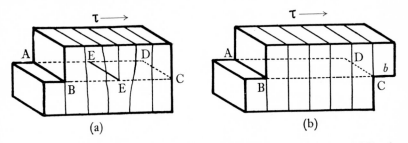

Fig. 4.10. (*a*) A simple form of dislocation on the slip plane ABCD of a block of crystal. (*b*) The movement of the dislocation to the surface has caused slip to occur over the entire plane.

slip takes place (fig. 4.10 b) by the movement of the dislocation line to the surface at CD, producing a step there as in fig. 4.4 b. When the dislocation EE passes across the slip plane of the crystal, the atoms behind it have sheared relative to the adjacent layer by the unit of slip, which is a definite amount and in a definite direction. The vector which defines this displacement is called the *Burgers vector* of the dislocation, and this is marked b in fig. 4.10 b. Experiment 4.6.2 (p. 82) describes how a bubble raft model of dislocations in crystals may be made.

4.4.4. *Work-hardening of metals*

It is convenient to define a *dislocation density* (ρ) as the total length of dislocation lines in unit volume of crystal. In undeformed metals a typical value of ρ would be 10^6 mm/mm^3, and this density can also be described as equivalent to the number of dislocations intersecting a unit area of the crystal (i.e. 10^6 mm^{-2}). Very much higher values of dislocation density can be observed in plastically deformed metals, so one must consider that dislocations are actually *produced* during plastic deformation. We therefore have to specify the presence of *dislocation sources* in crystals, and one type, known as the Frank–Read source, is illustrated in fig. 4.11. PQRS represents the slip plane of a crystal, and ABCD represents part of the network of dislocations within it. When a shear stress is applied to the crystal across the slip plane illustrated, the segment of dislocation BC starts to move, but the segments AB and CD do not, as they do not happen to lie on a slip plane. The dislocation BC is therefore effectively pinned at B and C, so the segment BC bows out. If the applied shear stress is high enough, the dislocation expands and passes around the pinning points B and C, forming a complete loop, the initial segment BC being

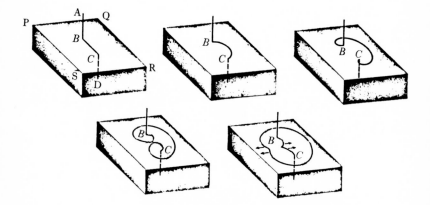

Fig. 4.11. A Frank–Read dislocation source : the dislocation ABCD is pinned at B and C. Under an applied stress BC bows out and generates a loop of dislocation—the segment BC being regenerated.

re-formed by a process of interaction and annihilation as in fig. 4.11. The re-formed segment BC can then repeat the process and create further dislocation loops on the slip plane.

When a metal is strained plastically, numerous sources of this type will operate. Dislocations which are moving on intersecting slip planes will interact with one another, will tend to become mutually entangled and therefore progressively more difficult to move. Thus ever-increasing stress is required to cause further strain, which is the phenomenon of *work-hardening* already referred to. The presence of dislocations can thus account for the softness of metals, and also for the fact that they become hard when deformed due to the dislocations meeting and obstructing one another's motion. Work-hardening is therefore associated with a progressive increase in the dislocation density of a metal, which can rise up to values of 10^{10} mm^{-2} at high deformations.

4.5. *The effect of high temperature*

4.5.1. *Recrystallization*

The dense array of tangled dislocations in cold-worked metals gives rise to a substantial *strain energy* stored in the crystal lattice (typically 5×10^7 J m^{-3}) so that the material is thermodynamically unstable relative to the undeformed state. It is, however, mechanically stable and can persist indefinitely at low temperatures. If the temperature is raised above 0.3–$0.5T_m$ (where T_m is the melting-point of the metal in °K) however, thermal energy can assist a very drastic softening process to occur, known as *recrystallization*. The effect is illustrated in fig. 4.12: new strain-free grains are produced by this treatment,

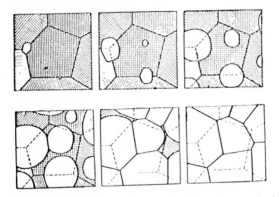

Fig. 4.12. Recrystallization: new strain-free grains progressively replacing a cold-worked structure (the dotted lines show the original grain structure).

which is called an *annealing* treatment. Grain boundaries of new grains sweep through the metal and replace the work-hardened grains by a new set of more perfect grains with junctions tending to 120° as explained in Chapter 2. This gives complete softening, with dislocation densities restored to values of the order 10^6 mm^{-2} once more.

The softening which takes place on annealing is taken advantage of in many commercial metal-working processes. If metals are shaped or forged by plastic forming in the temperature range *below* 0.3–$0.5T_m$, they will work-harden as the forging proceeds by the accumulation of dislocations within them. This type of forging is known as *cold working*. There is a limit to the amount of cold work to which a material may be subjected without danger of cracking or tearing, so that the metal is often annealed by heating above its recrystallization temperature for a time. The recrystallized material is then capable of further cold deformation.

If, for example, a large ingot is to be converted into a thin slab or sheet, such extreme changes of shape are usually carried out by forging *above* the recrystallization temperature. This is called *hot working*, and no work-hardening takes place at all under these conditions, since the material recrystallizes as it is being deformed, with the result that only comparatively small applied stresses are required. In hot-working operations, fine dimensional accuracy of the product is seldom possible, and furthermore there is a likelihood that the surface finish of the object will be poor, due to the oxidation of the metal during the process. Many forging operations therefore involve initial hot working in order to effect major shape changes with minimum expenditure of mechanical energy, then completion of the manufacture by a cold-working stage which can give a high dimensional accuracy, a good surface finish and, furthermore, a gain in strength due to work-hardening from the multiplication of dislocations in the material.

The rate of recrystallization depends on several factors—the amount of prior deformation, the temperature of anneal and the sample purity being the most important. A greater degree of cold work increases the rate of recrystallization at a given temperature, and Cottrell suggests that, since the driving force for moving the boundaries is provided by the energy of the dislocation lines, we can picture these dislocations as strings attached at various points to a boundary, each string pulling the boundary into its own grain. On one side of the boundary is the cold-worked grain which may have say 10^9 strings mm^{-2}, and on the other side is the recrystallized grain, which may only have around $10^4 mm^{-2}$, so the boundary moves into the worked grain at a rate which is proportional to the difference in dislocation density. Impurities are known to obstruct the motion of dislocations and of grain boundaries, so that the importance of impurity content upon recrystallization is clear.

When the growing crystals have consumed all the strained material and recrystallization is complete, *grain growth* takes place if the annealing is continued. Grain boundaries straighten, small grains shrink and the larger ones grow, so that the metal lowers its energy by reducing its total area of grain interface. A grain-boundary surface is a region of higher energy than a grain interior, because its atoms are less tightly bound than those within the body of the grain, and it is these ' surface-tension ' forces which cause the boundaries to migrate. An experimental analogy of this process is described in § 4.6.4 (p. 84).

4.5.2. *Creep*

If a metal is held under stress at temperatures at which diffusion rates are appreciable, dislocations may, over a period of time, be able to migrate within the metal crystals. This leads to an observed time-dependent deformation known as *creep*, which is of great practical importance.

Figure 4.13 shows the results obtained if the strain in a given sample of material under a constant load is measured over a period of time at a temperature in excess of $0.5T_m$. The curve shows an immediate strain 0A when the stress is applied, then *primary* creep AB which is a stage of decreasing creep rate, followed by *secondary* creep BC during which the creep rate is approximately constant and a *tertiary* stage CD during which the creep rate accelerates until the material fractures.

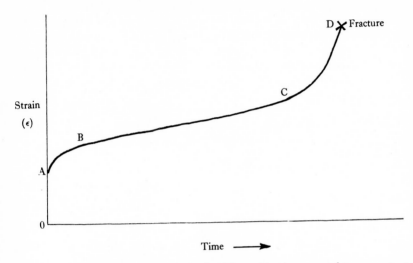

Fig. 4.13. A typical creep curve obtained by subjecting a metal to a constant load at elevated temperature.

At low stresses and temperatures only primary creep occurs and extension may eventually cease ; at high stresses and temperatures tertiary creep predominates—the acceleration is generally due to the growth of cracks in the material leading to a steadily rising stress and ultimate catastrophic failure.

The decrease in creep rate during the primary stage can be interpreted in terms of work-hardening making dislocation movement progressively more difficult. The steady-state secondary creep stage suggests that during this period the properties of the material remain constant, which would arise if the work-hardening occurring was balanced out by some thermal *softening* process. Creep strain usually involves other processes in addition to the movement of dislocations. For example, sliding at grain boundaries is known to occur under certain conditions—so that a fine-grained material would creep more rapidly than a coarse-grained one. Again, particularly at temperatures very close to the melting-point, solid-state diffusion (see Chapter 2) can contribute significantly to the process. A simple creep experiment is described in § 4.6.5. on p. 84.

81

Creep-resistant materials are essential in applications when stresses at high temperature have to be withstood without change of dimensions in service. The blades of gas turbines are an obvious example of this type, in that they may have to withstand high stresses when they are red hot. On the other hand, some metals of low melting-point can readily creep at room temperature : this is the reason why the lead covering of church roofs must be periodically replaced, and why unsupported lead pipes will gradually sag.

4.6. *Some simple experiments*

4.6.1. *Observation of slip bands*

It is easy to observe the slip bands on a bright surface of plastically deformed lead, if a low-power microscope (say $50\times$ magnification) is available.

If 0·02–0·03 kg of lead is melted in a crucible by heating over a hot bunsen flame, a suitable deformation specimen can be made by pouring the liquid metal onto a flat cold surface, such as a plate of smooth steel. When it is cold, examine the upper surface of the metal under the microscope by reflected light ; the individual metal grains will be visible due to the development of grain boundary grooves during the solidification process.

Now gently bend the lead through a small angle and examine the surface once more : a fine array of slip bands will be seen, whose orientation will change from grain to grain. Increase the amount of plastic deformation and note the increased density of slip bands ; as the lead is progressively bent, the onset of work-hardening will be detected as an increase in stiffness of the specimen.

4.6.2. *Bubble raft model of dislocations in crystals*

Simple lattice defects can be demonstrated by making a two-dimensional lattice of bubbles floating on a liquid surface, as first shown by Bragg and Nye. As bubbles become smaller they approximate to rigid spheres, and they can be made to represent the behaviour of a metallic structure because the bubbles are of one type only and are held together by a general capillary attraction which represents the binding force of the free electrons in the metal.

Take a medium-sized black plastic developing-dish and fill it with detergent to a depth of about 15 mm. Prepare a fine glass nozzle and blow bubbles of approximately 2 mm diameter with the nozzle held at a constant depth under the surface employing a steady stream of air at constant pressure. In this way bubbles of constant size are produced, and if the bubbles are prevented from piling up into more than one thickness by wafting them away from the source with a spatula, large rafts of bubbles may be produced.

(a) Grain boundaries

Single two-dimensional crystals of bubbles will be seen to form in close-packed planes which contain three directions at 120° to each other in which close-packed rows of bubbles exist. By uniting two misoriented crystal rafts of this kind, a ' bi-crystal ' can be made containing a grain boundary.

How localized is the misfit associated with the boundary? Estimate the width of the disturbed regions in terms of the number of ' atom spacings '. Compare your result with that observed in metals by field-ion microscopy (see fig. 2.15).

(b) Recrystallization

If a bubble raft is stirred with a glass rake and left to adjust itself, a process analogous to the annealing of deformed metals may be observed. The disturbed structure consists of many small grains (or ' sub-grains ') each of which contains defects in its structure.

Identify the processes whereby larger, more perfect grains are formed.

(c) Deformation behaviour

Prepare a single crystal bubble raft ; opposite edges of the raft may be deformed by shearing if two glass slides are held one in each hand, so that they just dip in the surface of the liquid and are moved so as to deform the raft.

By compressing or extending the raft, elastic deformation can be demonstrated. Observe whether the strain is homogeneous—i.e. if each ' atom ' of the crystal is displaced and returns to its original position when the strain is removed.

More extensive straining will cause plastic flow to take place. This occurs by a process of slip. Note the *direction* of slip.

The slip occurs, not by bodily slipping of one row past another, but by the movement of a *dislocation* along the slip direction.

Investigate the interaction of dislocations with (a) other dislocations, (b) vacant lattice sites and (c) grain boundaries.

N.B. Because the raft only contains one layer of bubbles, the length of the dislocation *line* (which runs perpendicular to the raft) is equal to one bubble diameter, of course, and it should be remembered that in real crystals dislocations are two-dimensional or ' linear ' faults, and not ' point ' defects as observed in the bubble model.

4.6.3. *Deformation twinning*

The audible clicks emitted by metals when they form mechanical twins are most easily demonstrated with tin.

About 0·05 kg of tin should be melted in a heat-resistant glass tube of about 5 mm diameter. After being allowed to solidify, the rod of metal may be extracted by carefully breaking the glass by crushing in a vice.

If the metal rod is now abruptly bent, a marked creaking sound will be heard as each individual grain forms bursts of deformation twins. The stresses to cause slip are higher in metals such as tin, which is not of cubic or hexagonal crystal structure, so that when strain is applied, yielding by twinning is preferred.

4.6.4. *A dynamic model of grain growth in metals*

A model due to C. S. Smith may be constructed which provides a most impressive analogy to the grain growth observed in metals when they are annealed at elevated temperatures for an extended period of time.

A glass tube of approximately 20 mm internal diameter and 200 mm in length should be sealed at one end. An aqueous soap solution or dilute liquid detergent should be added to give a depth of about 50 mm. The tube should now be sealed off under a rough vacuum to give a total length of some 150 mm : this is best done by connecting the tube to a water pump and boiling the liquid for a few minutes prior to sealing off the glass tube.

A fine froth of bubbles of a few millimetres' diameter can be produced by vigorously shaking the tube. Too good a vacuum makes it impossible to produce a froth—whereas froths at atmospheric pressure are very slow to change.

After producing a froth, hold the tube vertically so that the remaining liquid drains to the bottom of the tube and study the shapes of the individual bubbles of the froth. These have similar shapes to the grains of a polycrystalline metal ; this is particularly striking at the surface of the glass, where the shape of the bubble array is closely analogous to that observed in polished micro-sections of single-phase metals.

If the froth is studied for a period of time, a gradual increase in average bubble size of the froth will be observed. It is believed that this is geometrically, though not mechanically, a nearly exact model of grain growth as it occurs in metals. Estimate the average bubble diameter (d) at different times after the start of the experiment (t), and plot a graph of d against $\sqrt{t}$. Comment upon your result.

4.6.5. *The creep of metals*

Metals of low melting-point such as lead or solder, which is a lead–tin alloy, exhibit creep at and around room temperature. In the following experiment the form of the creep extension of a wire subjected to a constant tensile force is investigated as a function of temperature by a method due to Dr. G. W. Groves.

Equipment required

Lead wire or solder, about 1 mm diameter.
Metre rule.
Balance weights.
Tin can, rods.
Clock, thermometer, beaker, clamps.

Experimental details

It is convenient to apply the tensile force to the wire via a lever so as to reduce the size of the weights needed and to magnify the elongation. The suggested apparatus is sketched below. The lead wire is first attached to the gripping rods as indicated in fig. 4.14. The

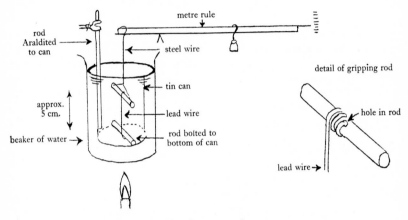

Fig. 4.14.   Sketch of suggested creep apparatus.

lower gripping rod is then bolted to the bottom of the can and finally the upper rod is hooked to the lever arm via a steel wire. The rest of the sketch is self-explanatory.

The maximum range of temperature which can be covered at one force is about 50°C, allowing for about 1 hour spent at the lowest temperature (slowest creep rate). The time for creep to failure at the highest temperature is then about 1 min. If the time available is more limited, the temperature range must be reduced so that the force giving fast enough creep at the lowest temperature does not give too rapid failure at the highest.

Start the experiment at the highest temperature, say 70°C, by finding a force which will produce failure in about 1 min. For a 1 mm diameter lead wire, this is about 6 N. Then repeat measurements of elongation at the same constant load at three lower temperatures. The region below room temperature can be explored with the use of ice and ice with salt if desired, using room temperature as the highest temperature and a force sufficient to produce failure in a short time at room temperature. The temperature should be held constant to within 2°C during a test.

Plot the elongation e (increase in length/original length) as a function of time for each test.

The elongation–time curves should show a region of roughly constant elongation rate in the middle of each test, with more rapid rates at the start and finish of a test. This is called the region of steady-state creep. Check whether the steady-state creep rate de/dt obeys a relationship of the form :

$$de/dt = A \exp - U/RT,$$

where A is a constant, for a given tensile stress, and R is the gas constant. If so, find the value of the constant U, called the activation energy per mole.

The activation energy for steady-state creep should be approximately equal to the activation energy for self-diffusion in the metal, i.e. the activation energy governing the random motion of the crystal's own atoms, from one site to another. This is because the mechanism of steady-state creep involves the migration of atoms through the crystal. Crudely speaking, this movement of atoms is needed to repair damage to the crystal produced by slip, and so to allow slip to continue.

Supplementary reading list

The Structure and Properties of Materials. Vol. III, Mechanical Behaviour : H. W. Hayden, W. G. Moffatt and J. Wulff (Wiley), Chapters 1 and 2, 5 and 6.

Metals Reference Book : C. J. Smithells (Butterworths) lists physical and mechanical properties of many metals and alloys.

CHAPTER 5

strong metals

WHAT do we mean when we say that a metal is strong? Do we mean that it is hard to deform it, or that it is hard to break it? In this chapter we will consider the first definition, and the final chapter will be concerned with the latter. Since we have some understanding of *how* metals deform, we are in a position to postulate general principles for increasing the resistance to deformation of metallic materials.

5.1. *The elastic modulus*

A high resistance to elastic deformation implies a high elastic modulus, and the value of this parameter is determined by the binding forces between the atoms of the crystal. In the field of non-metals, high elastic stiffness is found in structures in which all the atoms are joined by strong, covalent bonds or by directional bonds of more ionic character. The highest moduli are encountered in covalent solids with small atoms with small bond lengths, so that a large density of bonds per unit volume is produced. A covalence of three or four is required to ensure a three-dimensional network of bonds, and since the light elements of the Periodic Table principally form covalent bonds, the elements most likely to be found as constituents of crystals of high modulus are beryllium, boron, carbon, nitrogen, oxygen, aluminium and silicon. Certainly the *hardest* materials known are covalently linked solids : diamond (pure carbon) is the obvious example of this, but one could also quote corundum (aluminium oxide) and carborundum (silicon carbide) which are very important abrasives.

Among the metallic elements the transition metals are characterized by high moduli, and the tight bonding implicit in this effect is also reflected in the high melting-points associated with these elements (e.g. tungsten, niobium and molybdenum). As was observed in Chapter 4, the moduli of solid-solution alloys are usually intermediate between those of the constituent elements. The values of the moduli do not change markedly with changes in the metallurgical condition and microstructural state of the material, and so the elastic properties may be said to be 'structure-insensitive' properties. In fact, design engineers more usually select an alloy on the basis of its resistance to *plastic* yield, and it is this property which we will consider for the remainder of this chapter.

5.2. *Resistance to plastic flow*

The resistance of a metal to plastic flow may be defined simply by its elastic limit, or *yield stress* (fig. 4.3), but since this is a parameter which is difficult to measure, the *proof stress* is often quoted instead (see p. 70). Although the tensile test provides a fundamental measure of the resistance of a crystalline material to plastic deformation, it has the disadvantage of being a somewhat lengthy and therefore expensive test to carry out, requiring as it does a specially prepared and carefully machined test-piece.

A relatively quick and easy method of assessing the resistance of a metal to plastic deformation is the ' hardness test ', which measures its resistance to plastic *indentation*. A commercial hardness tester forces a small sphere, pyramid or cone of hard steel, tungsten carbide or diamond into a flat metal surface by means of an applied load. The hardness number (H) is defined as the load on the indenter divided by the area of contact between the indenter and the material, usually given in kg mm^{-2}. The hardness indentation itself causes plastic flow and therefore a certain amount of work-hardening to take place in its immediate locality, so that in general there is unlikely to be any simple relationship between the hardness number and the tensile yield stress (Y). However, in materials in which the rate of work-hardening is low (that is the slope of the stress–strain curve is shallow), to a good approximation it is found that

$$H \approx 3Y.$$

Since this is essentially a non-destructive mechanical test (the indentation being typically less than 1 mm in diameter in most forms of the test), coupled with its simplicity, it is frequently employed for quality control in industrial production processes.

Since plastic flow involves the motion of dislocation lines on the slip planes of crystals, the basic requirement for a material to be strongly resistant to plastic deformation is that dislocation movement should be obstructed. We can make it difficult for dislocations to move about in stressed metals by inserting into the microstructure of the crystal various *barriers* past which dislocations cannot readily migrate. In the case of pure metals, there are three general types of such barrier which we can consider :

(*a*) Point defects ; for example, lattice *vacancies*, where individual atoms are missing from the crystal pattern.

(*b*) Line defects ; for example, other dislocation lines threading the crystal.

(*c*) Planar defects ; for example, grain boundaries.

5.3. *Hardening by point defects*

Vacancies play an important role in the behaviour of metals at elevated temperatures, since, under those conditions, they can exist in equilibrium in the crystal and by their presence permit diffusion processes to take place easily. At lower temperatures, which concern us here, the importance of vacancies in influencing the behaviour of metals is not so fully understood, but there is good evidence that they can bring about a spectacular change in mechanical properties.

Two important ways by which point defects can be introduced into a metal are :

(i) By rapid quenching of the metal from a high temperature. This can be effective in retaining at the ordinary temperature of the quenching medium the concentration of vacancies which exists in

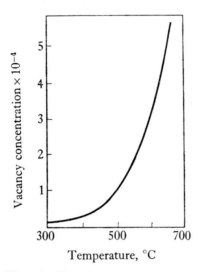

Fig. 5.1. The solubility curve for vacancies in aluminium.

thermal equilibrium at the higher temperature. The vacancy concentration in a crystal rises exponentially with increase in temperature (fig. 5.1 illustrates this for the case of aluminium) so that on quenching a large excess point defect population can be retained.

(ii) By bombardment of the metal with high-energy nuclear particles, such as by irradiation in a nuclear reactor. The high-energy neutrons can pass through the metal and occasionally collide with one of its atoms which then becomes a fast-moving ion in the crystal, having acquired sufficient energy to produce secondary ' knock-on ' damage in its vicinity.

The existence of excess point defects in a crystal gives rise to a hardening effect; the electron microscope has provided some evidence that the defects tend to cluster together, causing tiny ' knots ' of distortion in the crystal structure. When a moving dislocation encounters this elastic strain, extra stress has to be applied in order

89

to make the dislocation 'ride over' the defect, and this is reflected most prominently as an increase in the yield stress. Figure 5.2 illustrates the effect on the stress–strain curve of an aluminium crystal, (a) when slowly cooled, which would not contain an excess of defects,

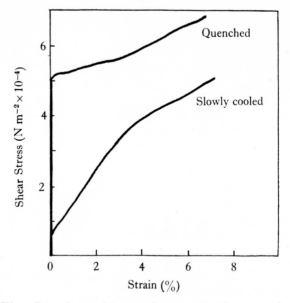

Fig. 5.2. The effect of quenching on the stress–strain curve of aluminium.

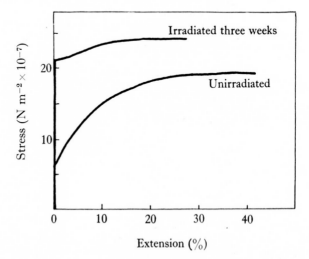

Fig. 5.3. Stress–strain curves for unirradiated and irradiated polycrystalline copper tested at 20°C.

and (*b*) when quenched from a high temperature, when a very pronounced increase in yield stress is apparent.

Figure 5.3 illustrates the phenomenon of 'radiation hardening' in fine-grained polycrystalline copper, tested at 20°C. The marked rise in yield stress arises again from the high concentration of excess point defects, and this factor is of critical importance in the selection of material by nuclear engineers for the construction of reactors, since the same effect can, in certain alloys, give rise to a serious embrittlement (see Chapter 6).

5.4. *Hardening by dislocations (work-hardening)*

As discussed on p. 69 and illustrated in fig. 4.3 plastic deformation progressively hardens a metal, due to the mutual obstruction of dislocations within the grains. Due to dislocation multiplication with strain (fig. 4.11), the dislocation density continually rises as deformation is continued, so that the number of interactions per unit volume between the dislocations will rise as the strain continues. An experimental demonstration of the effect is described in § 5.9.1 on p. 105.

Let us for simplicity consider the work-hardening of a single crystal —say in the form of a wire, by measuring the change in stress upon it as it is progressively strained in tension. The form of the curve depends upon the crystal structure of the specimen.

5.4.1. *Hexagonal crystals*

A hexagonal single crystal contains only *one* family of close-packed planes (fig. 1.7) upon which slip can easily occur. A typical stress–strain curve for such a crystal is shown in fig. 5.4, for magnesium at room temperature : after a small initial elastic extension, the crystal

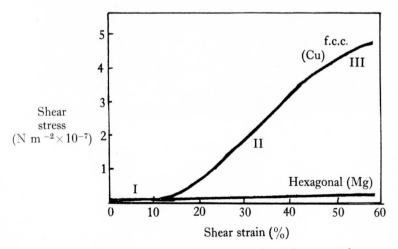

Fig. 5.4.   Stress–strain curves for hexagonal and f.c.c. crystals.

extends plastically (by the process illustrated in fig. 4.4) with only little work-hardening. The rate of work-hardening can be determined from the slope of this part of the curve, and for most hexagonal metals its value is approximately given by $10^{-4}\ \mu$, where μ is the shear modulus of the metal. Slip is only taking place in one direction in the set of slip planes (this is termed the 'operative slip system', and the dislocations moving on these planes mostly run out at the free surface and are lost, giving rise to the slip bands there, as already discussed (p. 70).

5.4.2. *Face-centred cubic crystals*

The shape of the stress–strain curve of f.c.c. crystals depends upon their orientation with respect to the axis of strain, but are more complicated in shape than those for h.c.p. metals. Figure 5.4 illustrates the behaviour of a copper crystal at room temperature: after an initial elastic strain, the plastic curve shows three regions or

Fig. 5.5. Electron micrograph of a thin foil taken from a crystal of plastically deformed aluminium showing dislocations tangled into ' cell ' walls.

stages, marked I, II and III. During stage I, the specimen is behaving like a hexagonal crystal, with dislocations moving only on one system, and the slope of the curve is again approximately $10^{-4}\mu$. This stage is called ' easy glide ', and is observed because *one* family only of close-packed planes experiences the maximum shear stress which enables slip to occur.

As the crystal stretches, its orientation changes with respect to the applied strain, and eventually *another family* of close-packed planes (of which there are four, fig. 1.8) experiences enough stress to start slipping. This will give rise to two sets of surface slip bands (fig. 4.5), and as soon as plastic deformation occurs on intersecting slip systems in this way, rapid work-hardening sets in. This corresponds to stage II of the curve in fig. 5.4, and the slope here is typically of the order $\mu/150$. The curve rises linearly, and then (at a stress that depends upon the temperature of test) the rate of work-hardening falls off (stage III, fig. 5.4) and the curve follows a roughly parabolic path.

Rapid work-hardening is apparently caused by the mutual obstruction of dislocations gliding on intersecting systems, and ample evidence of this process has been provided by the study of deformed metal foils in the electron microscope. The dislocations influence one another through their local elastic ' stress fields ', which are somewhat analogous to the magnetic fields surrounding a current-carrying conductor. The dislocations penetrate one another's slip planes, so a given dislocation will encounter these in its path like the trees in a forest, and they have to be cut through for glide to continue. This leads to the mutual entanglement and immobilization of dislocations on different slip systems, so that work-hardening is accompanied by the development of dense networks forming a rough cell-like structure in the crystals, as illustrated in the electron micrograph of fig. 5.5.

At the high stresses associated with stage III, dislocations can begin to *by-pass* the various obstacles, rather than accumulate at them, so that a *fall* in work-hardening rate is observed.

5.4.3. *Body-centred cubic crystals*

Crystals of this structure show a general pattern of behaviour very similar to that for f.c.c. crystals, although some of the detailed dislocation interactions are still imperfectly understood.

5.4.4. *Work-hardening of polycrystals*

When a polycrystalline specimen is stressed, each individual grain does *not* deform as if it were an unconstrained single crystal. Instead each grain is deformed into a shape that is dictated by the deformation of its neighbours, which requires the operation of *several* slip systems.

In cubic metals, therefore, intersecting glide and work-hardening begin immediately the plastic range is entered, and practically all the deformation in the individual crystals occurs under stage III conditions. The stress–strain curve (fig. 5.6) generally lies between the different curves which may arise from single crystals of extreme orientations— i.e. those whose slip planes are least favourably and most favourably oriented for slip with respect to the applied stress.

Because hexagonal crystals possess a more limited number of slip systems than cubic crystals, they are unable readily to deform under the constrains of a polycrystalline structure, and additional deformation processes, such as mechanical twinning, which require much higher stresses must operate. This results in hexagonal polycrystals

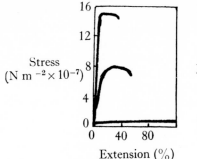

Stress
$(N\ m^{-2} \times 10^{-7})$

Hexagonal polycrystal (Zn)

F.C.C. polycrystal (Al)

Hexagonal single crystal (Zn)

Extension (%)

Fig. 5.6.   Comparison of single and polycrystal stress–strain curves.

being much harder to deform, and they show much less ductility than single crystals of these metals, as shown in fig. 5.6.

5.5. *Hardening by grain boundaries*

The preceding section has drawn attention to the constraining effect of grain boundaries upon plastic flow, and we will now consider the effect of the *grain size* itself upon the plastic properties of metals. A strengthening effect is observed in polycrystals, not only because

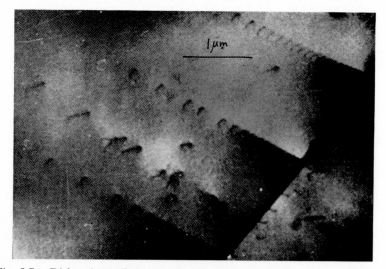

Fig. 5.7.   Dislocations piled up against a grain boundary in slightly deformed stainless steel.   (Courtesy P. B. Hirsch.)

94

of the complex intersecting slip process in the various grains, but also because the grain boundaries are themselves obstacles to the passage of dislocations. Hence slip cannot propagate freely from grain to grain, and fig. 5.7 illustrates this effect in a thin foil of lightly deformed stainless steel. A boundary between two crystals runs across the field of view, and dislocations moving on a number of parallel slip planes (appearing as short lines, since they run from the upper to the lower surface of the foil) in one of the grains are held up at the grain boundary due to the change in orientation. A 'traffic jam' of dislocation lines has built up near the interface.

When any polycrystal is stressed so that slip starts in one 'favourable' grain, the slip will tend to be blocked at the grain boundaries. A *stress concentration* is produced where the slip band in the first grain meets the neighbouring grain boundaries, which helps to 'trigger off' slip in these neighbouring grains. This effect eventually enables plastic yield to extend across the entire section of the specimen at a yield stress (σ_y) given by :

$$\sigma_y = \sigma_i + k_y\, d^{-1/2}.$$

This relationship is known as the Hall–Petch equation, and d is the grain size of the material, σ_i is the stress required to move dislocations

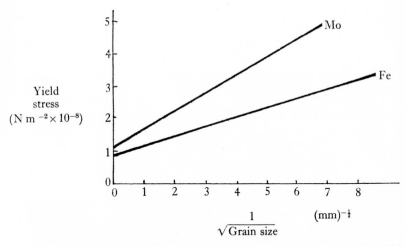

Fig. 5.8.   The dependence of the yield stress at room temperature in iron and in molybdenum upon grain size.

within the grains themselves and for a given metal k_y is a constant for a given temperature and rate of strain.

There is evidence that a number of metals obey the Hall–Petch relationship, and fig. 5.8 shows data for the yield stress of iron and of molybdenum as a function of grain size, plotted in accordance with it.

5.6. *Alloy-hardening*

As soon as we pass from considering pure metals to considering alloys, further new means of hardening present themselves. The presence of foreign atoms in the crystal can impede dislocation motion in several ways : we will consider the situation (*a*) when the alloying element is *soluble* in the metal and (*b*) when the addition of the alloying element gives rise to the presence of a *second phase* in the microstructure.

5.6.1. *Solid-solution hardening*

The effectiveness of foreign atoms as barriers to the movement of dislocations depends firstly upon the *size* difference between the solute atoms and the atoms of the parent crystal, and secondly upon

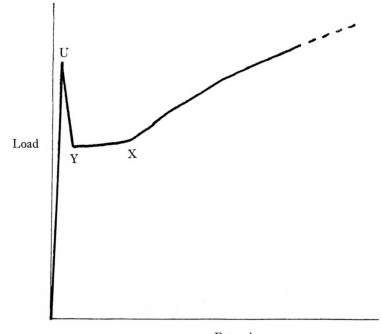

Fig. 5.9.   A large yield point in the stress–strain curve of a low-carbon steel.

the *proportion* of foreign atoms present in the crystal.   Elastic strain fields will be set up around the misfitting atoms, and these will have the effect of making dislocation motion more difficult.

Interstitial solute atoms in b.c.c. crystals provide a commonly encountered example of one important class of solute-hardening alloys. For example, carbon or nitrogen dissolved in iron produces an asymmetrical local strain which gives rise to a very steep rise in yield stress with increasing solute.   If only a very small quantity of such a solute is present, one effect of this high local strain is that the solute element will tend to migrate to the space provided along the dislocations present, rather than being uniformly distributed in solution in each crystal.   This has the effect of 'pinning' the dislocations in place, which not only has an interesting effect upon the shape of the initial part of the stress–strain curve of such an alloy, but also may be important technologically, as it is in the case of nitrogen in iron.

Only a very small quantity of, say, nitrogen in iron is needed to pin all the dislocations in this way.   We can estimate it as follows : 1 mm³ of metal typically contains about 10^6 mm of dislocation lines. Of the total number of atoms in this volume (1 mm³), say 4×10^{19}, the number lying on dislocations will be $10^6 \times$ the number of atoms per millimetre, say 4×10^6, i.e. 4×10^{12}.   The total solute content required to produce one foreign atom at every site along each dislocation is therefore

$$\frac{4 \times 10^{12}}{4 \times 10^{19}} \times 100, \quad \text{or } 10^{-5} \text{ atomic per cent of the element.}$$

Commercially 'pure' iron and low-carbon steel may well contain much more nitrogen than this, and its effect on pinning the dislocation lines can be detected by performing a tensile test, when a so-called *yield point* or sudden drop in load occurs when plastic flow sets in (fig. 5.9).   This yield drop can be felt if a piece of florist's iron wire is gently bent in the fingers.   The effect can be explained as follows :

The strain rate of a crystal is proportional to N (the number of dislocations moving per unit area and v (the average velocity of the dislocations).   In a tensile test a certain strain rate is applied, and N dislocations have to move at velocity v, but if the number of dislocations increases to say $10N$, the velocity needed is only $v/10$. If this lower velocity can be achieved with a lower stress, it follows that a drop in stress will occur.   Most of the dislocations in annealed iron are pinned by interstitially dissolved atoms, so that yield begins at a high stress (U, fig. 5.9), because the few mobile dislocations have to move rapidly.   These dislocations then quickly multiply (e.g. by the process illustrated in fig. 4.11) and the crystal acquires many new, unpinned dislocations, so N rises, v falls, and the stress for subsequent plastic flow falls (to Y, fig. 5.9) until (X, fig. 5.9) work-hardening causes it to rise again.   Over the reaction YX, known as the 'yield point elongation', the test-piece yields irregularly and bands (known as Lüder's bands) form on the surface which distinguish those parts of the specimen that have yielded from those which have not.

These Lüder's bands frequently form on the surface of steel sheets subjected to pressing and stamping operations.   Unsightly surface markings called 'stretcher strains' are formed which are avoided by slightly rolling the sheet just before pressing, which gives a uniform supply of unpinned dislocations, and thus avoids the incidence of the yield point. Interstitial carbon and nitrogen atoms can, however, diffuse in an iron crystal at moderately rapid rates at room temperature, and if a pre-rolled sheet is left overnight, for example, the dislocations will have become pinned once more by them, and the yield-point trouble will re-appear. This re-pinning effect is known as 'strain-ageing' of the steel.

Substitutional solute addition is the commonest way of solution-hardening a metal, and fig. 5.10 illustrates the effect of increasing

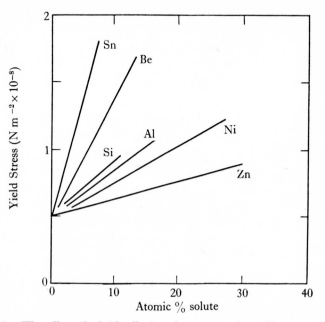

Fig. 5.10. The effect of soluble alloying elements on the yield stress of copper.

quantities of a number of solutes upon the yield shear stress of copper single crystals. It is seen that the shear stress–composition curves are linear, but the slope varies markedly from one alloy to another. This variation has been accounted for theoretically, and it arises from differences in modulus and of differences in atomic size between the solvent and solute atoms. Brass (copper–zinc) and bronze (copper–tin) are two familiar materials which make use of this hardening effect.

5.6.2. *Precipitation hardening*

As we discussed in Chapter 3, alloys which display a decreasing solid solubility with decreasing temperature (e.g. fig. 3.12) can be heat-treated to produce a microstructure containing finely dispersed precipitates.

These precipitates can have a profound effect upon the mobility of dislocations, and it is possible to produce large changes in the hardness of such alloys by suitable heat treatment, and this is the most important general method of strengthening metals. A great advantage is that the required strength can be induced in a product at the most convenient stage in its manufacture. For example, the alloy may be retained in a soft form throughout the period when it is being shaped by forging, by quenching it to form a single-phase, supersaturated

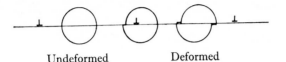

Undeformed   Deformed

Fig. 5.11. The shearing of a precipitate particle by a dislocation (diagrammatic).

solid solution. Forging is now cheap and easy to carry out, and the material is finally hardened by precipitation, so that in service it will have a high strength.

On being held up by a particle, a dislocation can continue in its path across the crystal in two possible ways. If the particles are very close together, the dislocation may *cut through*, as shown schematically in fig. 5.11, and the particles are seen to be sheared with the rest of the crystal. If the particles are further apart, *looping between* the

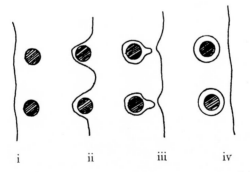

i   ii   iii   iv

Fig. 5.12. Plan view of a slip plane with particles and dislocation. The dislocation loops between the particles, leaving a ring of dislocation trapped at the particles.

particles may take place by a process which is represented in a simple way in fig. 5.12. In this case the particles are not sheared, and some dislocation ' debris ', in the form of small loops at the particles, is left behind when the dislocation has gone past.

The shape of the stress–strain curves to be expected from these two mechanisms are shown in fig. 5.13. Curve S represents the behaviour of the single-phase material, and curve C that when a cutting-through process obtains. The general shapes of the two curves are similar, but the stress level is raised in case C, due to the difficulty of forcing the dislocations through the precipitates. Curve L indicates the type of behaviour when dislocation looping takes place: a much

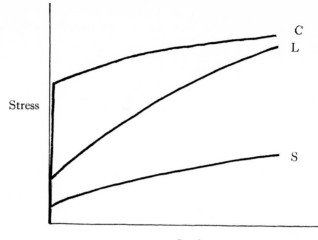

Fig. 5.13. Stress–strain curves of: S, single-phase material, C, two-phase alloy in which particle-cutting occurs, L, two-phase alloy in which dislocation looping occurs.

greater rate of work-hardening is observed, arising from the more rapid build-up of dislocations in the structure as strain proceeds, due to the ' debris ' left by the looping process.

When a suitable alloy is quenched and then aged to produce a precipitate, the second phase first forms in a very fine dispersion, but as ageing proceeds the particles gradually increase in *size*, so that the average *spacing* between the particles also increases.

As the particles grow, the shear stress increment ($\Delta\tau$) required to make the dislocations cut them also rises (curve C in fig. 5.14), whereas the stress increment required to cause dislocation looping (curve L, fig. 5.14) decreases as the inter-particle spacing increases. As ageing continues, the measured yield stress would therefore be expected to follow the form of the dotted curve in fig. 5.14, and this general pattern of behaviour with an optimum ageing time to give a maximum hardness is commonly observed in many commercial alloys. The time to peak hardness depends on the solute diffusion rate in the metal, and thus on the particular ageing temperature : for periods longer than this the hardness falls and the alloy is said to be ' over-aged '.

100

There are a number of aluminium-based alloys which can be precipitation-hardened—both wrought and cast products are manufactured and subjected to this treatment. The Al–Cu, Al–Mg–Si, Al–Mg–Cu and Al–Mg–Zn are the most important families of such alloys.

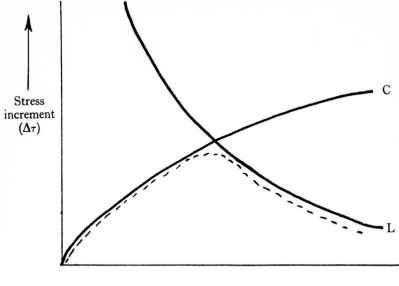

Fig. 5.14. The change in yield stress with ageing time in a precipitation-hardening alloy (dotted curve).

Copper-based alloys containing about 2 wt. % Be are also hardenable in this way to produce the strongest and hardest of the known copper alloys.

5.7. *Strength at high temperatures*

So far we have only considered the strengthening of metals for use at ordinary temperatures, and our ideas will have to be modified when the metal is subjected to stresses at temperatures above $0.3–0.5T_{\mathrm{m}}$ (where T_{m} is the melting-point, in $^{\circ}$K). Under these conditions thermal energy can assist atoms to diffuse at appreciable rates within the solid, and it is found that the barriers to dislocation movement described above become less effective, so that metals and alloys become softer. We will consider in turn the various barriers to dislocations.

5.7.1. *Point defects (vacancies)*

At elevated temperatures any excess vacancies can readily migrate and can be absorbed in the metal at grain boundaries or dislocations or lost at the metal surface. Thus any vacancy hardening (figs. 5.2, 5.3) rapidly anneals out under these conditions.

101

5.7.2. *Line defects (dislocations)*

When work-hardened material is heated in this temperature range, the very drastic softening process of *recrystallization* takes place (fig. 4.13) so this method of hardening is also clearly inappropriate for conferring high temperature strength.

5.7.3. *Planar defects (grain boundaries)*

If a fine-grained metal is held at elevated temperatures, it is observed that a general grain-coarsening takes place by the migration of grain boundaries. The driving-force for this grain growth has been discussed on p. 80, and this effect will clearly lead to some fall in strength, as implied by the graph of fig. 5.8 and by the Hall–Petch equation which it illustrates.

5.7.4. *Alloy hardening*

The effect of solute elements upon mechanical properties usually declines as the temperature is raised. In the case of dislocations pinned by solute atoms, thermal energy can assist in the unpinning

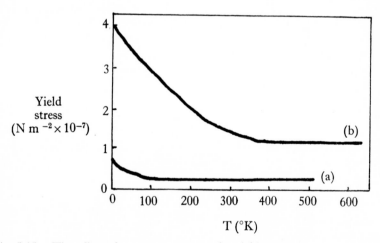

Fig. 5.15.   The effect of temperature upon the yield stress of (a) pure copper and (b) of a copper–zinc solid solution.

process, so that yield-point effects disappear at high temperature. Ordinary substitutional solute hardening also falls off as the temperature rises, although the evidence suggests that the effect does not entirely disappear at the highest temperatures, as show in fig. 5.15 which shows the change in shear yield stress for copper crystals and for crystals of copper–zinc solid solutions.

In two-phase alloys the solubility of foreign atoms increases as the temperature is raised so that particles of second phase will tend to

102

dissolve in the crystal in which they are dispersed. There will also be a tendency for insoluble particles to agglomerate, so that fewer, coarser particles remain—thus precipitated alloys will over-age, and the large inter-particle spacing will be associated with a fall in yield stress, as predicted by the curves of fig. 5.14.

5.7.5. *Creep of alloys*

The implication of these elevated temperature effects is that if a metal is held under stress under these conditions, the dislocation barriers which existed at ordinary temperatures may be no longer effective so that, over a period of time, the dislocations may be able to migrate which is observable as the phenomenon of *creep* (§ 4.5.2).

The problem facing the designer of alloys which will resist stresses at high temperatures is thus one of placing barriers which are thermally stable to prevent the movement of dislocations. Many such alloys feature a fine dispersion of hard particles which are almost insoluble in the metal and which both impede the movement of dislocations within the grains and also reduce the tendency of the metal to soften by recrystallization or coarsening of the grains. Although the structures of today's most widely used alloys for resistance to creep (for example, those chosen to manufacture turbine blades for jet engines) are usually developed by solution-treatment, quenching and precipitation heat treatments, the techniques of *powder metallurgy* are also used to produce materials with spectacular strengths at high temperatures.

The simplest application of the technique is simply to blend together finely divided powders of the metal and a hard stable compound. The mixture is compacted under pressure in a die and heated ('sintered') to increase its density. It is difficult in practice to prevent the hard particles from clustering during pressing and sintering, which causes a drop in hardness, and an important step forward was made with the development of Sintered Aluminium Powder (SAP) products, which possess excellent strength and outstanding stability at high temperatures. At 400°C this material has five times the strength of conventional aluminium alloys, and does not lose its strength even after long-term heating at temperatures of about 500°C. This material owes its strength to the fact that each particle of aluminium powder from which it is prepared has a film of aluminium oxide (alumina) upon its surface. When the compacted powder is manufactured, fine particles of alumina (which may constitute up to 15 wt. % of the alloy) become dispersed throughout the metal and, since alumina is a very stable high-melting oxide, make the structure stable.

More recently this same principle has been used to produce dispersion-strengthened lead by powder metallurgy. This time lead oxide from the surface of the original particles of the powder is dispersed throughout the material and sets as a stable barrier in the structure. Lead of this type has strengths between two and three times greater than that of pure lead, and since this material does not deform by creep at ordinary temperatures (by contrast with pure lead), there is scope for applying it to lead/acid battery grids, cable sheaths, chemical engineering and radiation shielding.

A true high-temperature material of this type has been developed in the U.S.A. in the form of nickel containing finely divided particles of thorium oxide (thoria). This was done by producing a very uniform mixture of fine particles of the oxides of nickel and thorium and reducing the nickel oxide to metal. Metallurgists have devised means of fabricating nickel–thoria powder into bars and sheets, and the resulting material—TD Nickel—is now in commercial production. The strength of TD Nickel is modest at room temperature, but it persists nearly to the melting-point of nickel, in analogy with the strength of SAP, so that in the range between about 1100°C and 1350°C, TD Nickel is stronger than any commercially produced alloy based on iron, cobalt and nickel. Furthermore, it has a lower density and greater resistance to oxidation than alloys based on the higher melting-point metals such as molybdenum (m.p. 2620°C) and niobium (m.p. 2468°C).

5.8. *Composite materials*

As we discussed in § 5.1, the hardest materials known have their atoms linked by strong covalent bonds, and to ensure a large value of the elastic modulus a large density of bonds per unit volume is required. As we discussed earlier, the elements possessing these properties are Be, B, C, N, O, Al and Si, and very hard materials always contain one of these elements and frequently only these (see pp. 68 and 87). The strong interatomic binding leads, furthermore, to very high melting-points in these covalent solids (many of them are greater than 3000°C), and the directional nature of the bonds implies non-close-packed crystal structures so that the densities of these materials will be low ; these are both attractive properties for the engineer.

Such solids are not used in structures because of their *fragility*, however, which arises because of their inherent brittleness (i.e. it is difficult to move dislocations in these crystals) and also from the fact that they usually contain small surface *cracks* or steps. These faults act as notches which cause a great local magnification of any applied stress, which is sufficient to break the bonds at the tip of the crack so that it can extend and run through the material.

If cracks and notches are eliminated from these materials they become extremely strong, but in practice cracks often arise through chemical attack or scratching and their strength is immediately lost. One way in which the high strength of covalent solids can be exploited is in the production of *composite materials*. The strong solid is produced in the form of long fibres, in which all but the smallest cracks are eliminated, and a composite material is manufactured consisting of a large number of parallel fibres bound together by a weaker, yielding matrix.

A composite of this type is extremely strong in tension parallel to the axis of the fibres, because the matrix prevents a crack from spreading from one fibre to the next. The matrix thus acts as a binder for the fibres, and also prevents the fibres from rubbing against one another and causing weakening cracks. A resin or polymer, or a ductile metal makes a suitable matrix material. Fibre-glass is a material using an organic binder which is in regular use today, although glass fibres have a lower modulus than fibres of asbestos or carbon and composites based on these latter fibres have much greater stiffness. Organically bonded composites cannot be used at high temperatures, and metallic matrices have to be used under these conditions. Figure 5.16 illustrates the superior properties that have

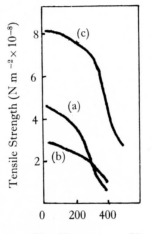

Fig. 5.16. A comparison of the strengthening obtained at elevated temperatures by (*a*) a conventional precipitation-hardening alloy, RR58, (*b*) sintered aluminium powder (SAP), (*c*) a fibre-reinforced aluminium composite.

been obtained from experimental materials designed on this principle of fibre reinforcement.

5.9. *Experimental*

5.9.1. *Cold working*

If a metal is plastically deformed at temperatures below which rapid diffusion can occur (e.g. below say $0.3T_{\mathrm{m}}$), the dislocation density within the metal progressively rises and *work-hardening* takes place. This process, called cold working, can readily be demonstrated.

Take a rod of copper (say of diameter 5 mm and length 110 mm), ensuring that it is initially in the fully annealed state by heating it in a bunsen flame to red heat (use tongs!) for a few minutes. If a length of 33 mm at one end of the rod is, after cooling, hammered flat while held on an anvil or against a heavy vice, the increase in the strength brought about by this treatment can be recognized by making a single scratch across the rod with the edge of a file. Repeat this (trying to apply the same pressure on the file each time) on the unhammered end of the rod. Examine the scratches closely with a magnifying glass, and note which is the deeper. Hold one end of the rod between the thumbs and forefingers and bend it slightly ; do the same at the other end of the rod ; which end seems the stiffer ?

5.9.2. *Annealing*

Now hold the hammered end of the copper rod in a bunsen flame until it is at a dull red heat (tongs!), and hold it there for 30 sec. Quench the hot end in water, so that the oxide scale is cracked and removed.

Compare the stiffness of the flattened and the original end of the rod now. Next perform the simple scratch test with the file and compare the hardnesses of the two ends of the rod. The changes in metallurgical structure which have taken place are illustrated in fig. 4.12.

5.9.3. *Hardening of steel (see § 3.3.5)*

Take a rod of high-carbon steel (i.e. containing approximately 0·8 wt. % carbon) about 3–5 mm in diameter and of length, say, 150 mm. Such rods can readily be obtained from a stockist of engineering materials, the alloy being known commercially as ' silver steel '.

In the as-received state the alloy consists of grains of ferrite (b.c.c. iron) containing relatively large particles (i.e. in excess of 10 μm diameter) of iron carbide, Fe_3C. The rods are therefore relatively soft and ductile : bend one as far as you can.

Now heat the middle of one rod in the centre of a hot bunsen flame until it glows a bright red. Under these conditions the iron is converted into the f.c.c. form (austenite) which has a high solubility for carbon (see fig. 3.15), and so all the particles of Fe_3C will dissolve and the single-phase solid solution is produced.

When the rod has been held for $\frac{1}{2}$–1 min at high temperature, quench it into water, and this rapid cooling causes a diffusionless transformation to take place : the austenite grains are converted almost instantaneously into a fine structure of *martensite* which is extremely hard due to its state of high internal strain. Try scratching the quenched end of the rod with a file in order to assess its hardness. Now try to bend the rod. What happens?

Now take a third rod, heat its centre as before and quench it as before to form martensite. Now hold the martensitic section of the rod about 150 mm above a cool bunsen flame and hold it there for several minutes. (The correct heat treatment can be assessed by abrading the rod surface with emery paper until it is bright and shiny, and then holding it above the flame as described until the oxide film formed upon it causes it to have a golden-yellow colour. Can you explain how the thickness of the film influences its colour.

The steel is now said to be *tempered* (p. 62), and the structure consists of b.c.c. ferrite once more, but with the carbide now re-precipitated in an extremely finely divided form, so that the inter-particle spacing is also very small. The particle size will be of the order 10^{-1} μm, and the spacing between the particles perhaps 1 μm, so that dislocation motion is severely impeded. Try

> (a) the scratch test with the file,
> (b) bending the rod.

With a fourth specimen, quench it to martensite, then temper at a dull red heat for a few minutes. Repeat the above tests. Can you account for the differences between the mechanical properties of the third and fourth specimens? (Note: at the higher tempering temperature, diffusion rates will be greater. What will be the effect of this on the precipitate particle size, for a given tempering *time*?)

Supplementary reading list

Introduction to Strengthening Mechanisms: D. K. Felbeck (Prentice Hall).
Hard Materials: A. Kelly (Wykeham Publications).

CHAPTER 6

fracture

PROVIDED the temperature is not too low, most metals are ductile and can be deformed to a considerable extent before they fracture. In service, however, many metallic materials have been found to fail in an apparently brittle way, leading to catastrophic damage in structures such as bridges and ships' hulls, and the phenomenon of fatigue failure is also all too frequently met in hulls and wings of aircraft, for example. In all cases the process of fracture may be considered firstly in terms of the *nucleation* of one or more microcracks within the material, followed by their *propagation* through the structure until total failure occurs. We will examine the various possible mechanisms whereby metals can fail in terms of these two stages.

6.1. *Ductile fracture*

It is essential in metal-working operations, such as forging or pressing, that the material should exhibit a high degree of ductility, so that it is important to understand the processes leading to fracture in ductile materials. The details of these processes have in fact only been fully understood comparatively recently.

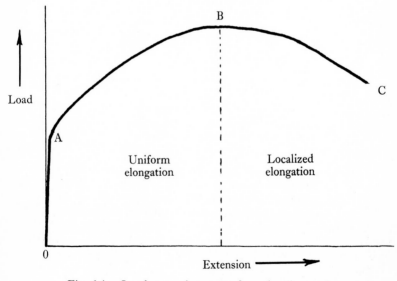

Fig. 6.1. Load–extension curve for a ductile metal.

108

If a test-piece of ductile polycrystalline metal in the form of a cylinder is subjected to tensile elongation until it fractures, the variation in load upon it will be typically as shown in fig. 6.1.   After an initial elastic strain 0A, the test-piece will plastically elongate uniformly over the region AB, the load rising due to work-hardening. As the material is extended further (BC), the strain becomes *localized* (fig. 6.2) and a constriction or ' neck ' forms in the test-piece ;   the

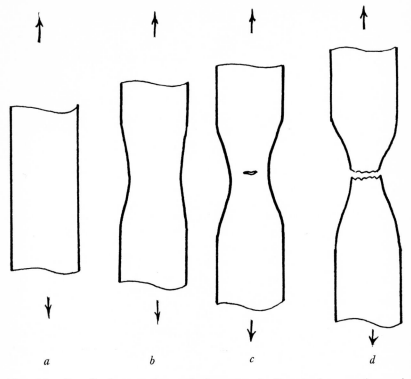

Fig. 6.2.   Ductile fracture by neck formation leading to ' cup and cone ' fracture.

neck deepens progressively as the material is deformed until, at C, the specimen finally fractures to produce a fracture surface which is approximately normal to the specimen axis, but which has a rim at approximately 45° to the axis—often described as a ' cup-and-cone ' fracture profile, as illustrated in fig. 6.2 *d*.   Although the metal in the necked region continues to work-harden as it is deformed from B to C (fig. 6.1), the progressive local reduction in cross-sectional area leads to a gradual fall in the load supported by the specimen as shown in the graph.   By dividing the value of the maximum load achieved during the tensile test by the *original* cross-sectional area of the test-piece, an engineering parameter known as the ' ultimate tensile stress ', or ' ultimate strength ' of the material is obtained.

A simpler form of ductile rupture occurs in some hexagonal metal single crystals, particularly if they are deformed at elevated temperatures. The process illustrated in fig. 4.4 can continue until final separation occurs by *shear* : zinc and cadmium have been observed to fail in this way, by a simple sliding apart on their slip planes. The fracture of ductile f.c.c. metal crystals is usually preceded by necking. This neck in pure metals may continue to develop until the cross-section is reduced to a point or chisel-edge, and the specimen separates to give this form of fracture at an extremely low final load.

6.1.1. *The mechanism of nucleation and propagation of ductile fracture in polycrystals*

If a tensile test is interrupted when a neck has developed in the test-piece, but before final fracture has occurred (i.e. between B and C in fig. 6.1), and the specimen is removed, sectioned longitudinally and examined metallographically, small pores are found to have developed in the necked region. The pores mainly originate at small

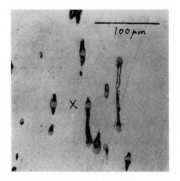

Fig. 6.3. Illustrating the role of inclusions in the nucleation of ductile fracture. Regions such as × subsequently fail by ' internal necking '. (Courtesy C. D. T. Minton and G. C. Smith.)

inclusions or particles of second phase, and appear to arise from the metal being pulled away at the interfaces of the particles as illustrated in fig. 6.3. For example, in the neck of copper tensile specimens, small cavities have been found to be associated with inclusions of copper oxide particles, which are commonly found in copper of commercial purity. As the purity of a metal is raised, inclusions of this type will be less numerous in the structure, so that the initiation of ductile fracture will be more difficult. A pure metal, free of inclusions, should thus ideally neck down to a point—and this has been found to be the case in some very pure materials.

The propagation of the ductile crack takes place by the linking-up of the pores as the neck is elongated. This coalescence can be regarded as a kind of 'internal necking' and fracture of the bridging regions such as that marked × in fig. 6.3. These elements of metal will behave like miniature tensile test-pieces and will stretch until they are drawn to a point and fail, thus linking adjacent pores. Figure 6.4

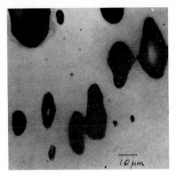

Fig. 6.4. Voids linking up in the necked region of a polycrystalline copper specimen. (Courtesy C. D. T. Minton and G. C. Smith.)

shows the microstructure within the neck of a copper specimen containing oxide inclusions, and the inter-linking of voids in this manner can be clearly seen. When a large internal cavity has formed roughly perpendicularly to the axis of the specimen, the final rim of metal will experience a high local stress and it will fail by shear at approximately 45° to the specimen axis, leading to the 'cup-and-cone' type of fracture of fig. 6.2. The flat part of the fracture, arising from the interlinking of voids, has a characteristic dull, fibrous appearance to the unaided eye, whereas the 45° rim, forming the 'ears' of the fracture surface, is fairly bright and featureless. Section 6.7.1 (a) on p. 129 describes a simple experiment to demonstrate ductile fracture in a solid.

6.1.2. The energy of fracture

An important property of the stress–strain curve is that the area under the curve gives the work/unit volume on the deformed material necessary to deform it to fracture. Consider a tensile specimen of length l, cross-sectional area A (volume $V = A \times l$) given an incremental extension dl by a force F, then the incremental work done (dW) is:

$$dW = F \times dl.$$

111

If the stress is σ we can write :

$$dW = \sigma \times A \times dl$$

$$= \sigma \times V \times \frac{dl}{l},$$

$$\therefore W = V \int_0^\varepsilon \sigma \, d\varepsilon,$$

where $d\varepsilon$ is the incremental strain. Then the work per unit volume is :

$$\frac{W}{V} = \int_0^\varepsilon \sigma \, d\varepsilon, \qquad . \qquad . \qquad . \quad (6.1)$$

which represents the area under the stress–strain curve.

In a ductile metal this area will be large, so that considerable energy is expended in fracturing the material. In other words, the greater the area under the stress–strain curve, the 'tougher' the material is said to be.

6.2. *Brittle fracture*

The problem of brittle fracture is of great importance. Many solids, particularly metals of b.c.c. cubic crystal structure and ionic solids, fracture in a brittle fashion at low temperatures, after only a small amount of plastic deformation, with a rapidly propagating crack, so that brittle materials are characterized by a low energy absorption to failure.

The path of a brittle crack is generally *across* the grains of the microstructure (i.e. is 'transgranular'), and lies parallel to certain definite crystallographic planes of the material, known as the *cleavage planes*. In the case of a few alloys (never in pure metals) *intergranular* cracking is encountered, and we will consider the latter situation separately later.

6.2.1. *The theoretical cleavage strength of a brittle solid*

The stress required to cause brittle failure along the cleavage planes of a crystal is known as the 'cleavage strength' of a crystal, and its value can be estimated theoretically. Consider a brittle crystal of unit cross-sectional area subjected to a tensile force, as the cleavage planes are pulled apart, the interatomic bonds are stretched and they thus accumulate energy. When fracture occurs, this stored energy (i.e. total work done) reappears as the *surface energy* of the two cleaved fracture faces. Since a brittle solid deforms only elastically until it fractures the stress–strain curve is linear, and the work done per unit volume in elastically straining the crystal $= \frac{1}{2}\sigma \times \varepsilon$ (i.e. area under this straight line). If $E = $ Young's modulus, we can write $\varepsilon = \sigma/E$, so, work done per unit volume $= \sigma^2/2E$. Considering the energy stored in the volume between adjacent atomic planes of spacing a, since the crystal is of unit cross-sectional area, the work done $=$ stored energy $= \sigma^2 a/2E$.

After fracture, the area of each new fracture face is unity, and if the surface energy per unit area is γ, we can equate these two energies, i.e.

$$\frac{\sigma^2 a}{2E} = 2\gamma,$$

so

$$\sigma = 2\sqrt{\left(\frac{\gamma E}{a}\right)}.$$

σ is the upper limit to the ideal strength of a crystal, since we have assumed Hooke's law to hold up to fracture. More realistically a sine or a Morse function can be used to describe the variation of σ with strain, and these are found to give a value about half that obtained above, so we can say that the theoretical cleavage strength is approximately :

$$\sigma_{\text{th}} = \sqrt{\left(\frac{\gamma E}{a}\right)}. \qquad . \qquad . \qquad . \quad (6.2)$$

6.2.2. *Brittle strength of amorphous solids*

In non-crystalline brittle material such as glass the fracture stress is three or four orders of magnitude lower than the theoretical strength estimated above. These relatively small forces for fracture are due to the presence of a flaw or crack which acts as a stress concentrator and leads to localized stresses of the order of the ideal strength. The strength of freshly drawn glass fibres approaches the theoretical strength, but rubbing with a finger is often sufficient to cause the fracture stress to fall catastrophically due to the creation of tiny surface cracks.

Figure 6.5 illustrates a cleavage crack of length $2c$ in a brittle material of unit thickness perpendicular to the diagram, held at a constant stress σ. For the crack to propagate, the maximum stress across the tip of the crack must be equal to the ideal cleavage strength (σ_{th}), and the energy necessary to advance the crack must have come from the elastic strain energy stored in the stressed specimen.

In fig. 6.5 a roughly circular region of radius c about the centre of the crack is released from stress and hence relieved of strain energy by the presence of the crack. If the stress–strain curve is linear, we can write :

$$\text{energy per unit volume} = \frac{\sigma^2}{2E},$$

so

$$\text{the energy relieved by the crack} = \frac{\pi c^2 \sigma^2}{2E}$$

per unit thickness perpendicular to the diagram. The crack can grow if the rate of release of elastic energy is at least as large as the rate of increase of surface energy required to form the crack face. Since there are two growing surfaces at each end of the crack, the crack will grow if

$$\frac{d}{dc}\left(4\gamma c - \frac{\pi c^2 \sigma^2}{2E}\right) = 0,$$

i.e.

$$4\gamma = \frac{\pi c \sigma^2}{E},$$

so

$$\sigma = 2\sqrt{\left(\frac{\gamma E}{c}\right)};$$

more exact calculations give :

$$\sigma = \sqrt{\left(\frac{\gamma E}{c}\right)}, \qquad . \qquad . \qquad . \quad (6.3)$$

which is the famous formula of A. A. Griffith.

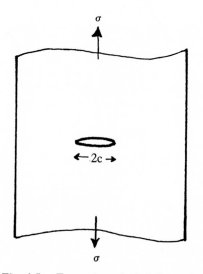

Fig. 6.5. Formation of a brittle crack.

Comparisons of equations (6.2) and (6.3) will indicate why most brittle materials are fragile and break at low stresses, since everyday objects contain cracks or flaws for which $c > a$ (the distance between the atoms). Inserting typical values of σ, γ and E for glass, values of c of the order 2×10^{-3} mm are obtained, which are close to observed crack sizes. Glass for windscreens is ' toughened ' by a heat treatment which produces compressive stresses in the surface which oppose the growth of such surface cracks.

6.2.3. *Brittle strength of crystalline solids*

The Griffith theory only considers the propagation of cracks in a material which does not undergo plastic flow prior to fracture. In crystalline solids it is recognized that plastic deformation is needed to nucleate a crack and that the propagation of the crack is accompanied by further plastic flow. This flow tends to blunt the crack, so that more energy is needed to make it propagate. This energy of plastic work P is in many cases so much greater than γ that the latter may be neglected, and the Griffith criterion now becomes :

$$\sigma = \sqrt{\left(\frac{PE}{c}\right)}. \qquad . \qquad . \qquad . \quad (6.4)$$

Section 6.7.2 on p. 130 describes a simple experiment to estimate the magnitude of P (equation (6.4)) in aluminium.

6.2.4. *The nucleation and propagation of cleavage cracks in metals*

(*a*) *Nucleation*

Dislocation interactions are responsible for the formation of crack nuclei in crystals, although the extent of dislocation movement prior to the propagation of a crack can be very small. For example, if

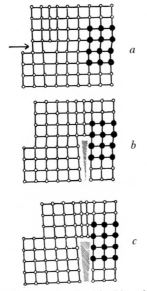

Fig. 6.6. Crack formation by blocking dislocations.

dislocation motion is blocked by a hard particle inside a crystal, or at a grain boundary, such a material may become vulnerable to crack formation through the simple mechanism illustrated in fig. 6.6. A

115

high concentration of stress collects where the dislocation is blocked (fig. 6.6 *b*), which by the coalescence of several dislocations (fig. 6.6 *c*) may nucleate a tiny crack.

Another process involves the movement towards each other of dislocations on two intersecting slip planes : where the dislocations

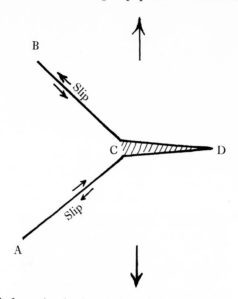

Fig. 6.7. Crack formation by interaction of dislocations on intersecting slip planes. The combination at C of dislocations arriving from B (on the slip plane BC) and from A (on the slip plane AC) can lead to the nucleation of a crack in the plane CD.

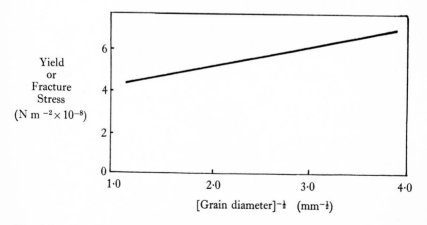

Fig. 6.8. Equality of fracture stress in tension and yield stress in compression for mild-steel specimens of different grain sizes at 77°K.

116

meet, they are able to combine to form a crack nucleus by the process illustrated in fig. 6.7.

One elegant demonstration that some type of plastic yield and dislocation interaction is involved in the fracture by cleavage of iron at 77°K, was to compare the *yield stress* under compressive stress (when fracture will not occur) with the *cleavage stress* under tensile stress. This was done on a number of specimens of different grain size, and it was shown (fig. 6.8) that the variation with grain size, both of the yield stress and the fracture stress, obeys a Hall–Petch relation, so that the mechanism controlling the propagation of yielding also controls the onset of fracture.

(b) Propagation

The crack nuclei spread across the cleavage planes of the grain, and a smooth, brightly reflecting fracture surface is produced. On a microscopic scale these bright cleavage facets are seen to have sharp steps on them which run roughly parallel to the direction of propagation of the crack front. These so-called ' river markings ' (fig. 6.9 a)

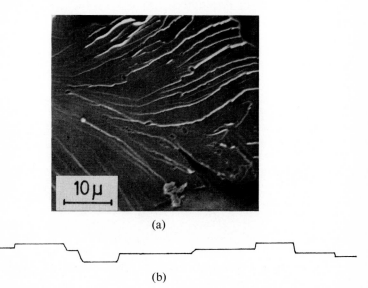

(a)

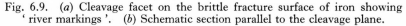

(b)

Fig. 6.9. (a) Cleavage facet on the brittle fracture surface of iron showing ' river markings '. (b) Schematic section parallel to the cleavage plane.

are highly characteristic of cleavage fracture and they arise where cracks running on parallel planes in the grain (but at different heights) link up and form a step which appears as a dark line under the microscope (fig. 6.9 b). Section 6.7.1 (b) on p. 129 describes a simple experiment to demonstrate cleavage fracture in zinc.

117

6.2.5. *Intergranular fracture*

In some alloys it is found that a crack nucleus can propagate more readily around the grain boundaries of the structure than across the cleavage planes, and this gives rise to an intergranular fracture, and it may be attributed either to the effect of impurity elements which have segregated preferentially to the grain boundaries and weakened them, or to the formation of a brittle film of a second phase. Either of these effects may permit ready crack propagation there. This mode of fracture has been observed in tungsten and molybdenum alloys containing oxygen, carbon or nitrogen, copper containing bismuth or antimony, and iron containing phosphorus, so that careful metallurgical control of these impurity elements has to be maintained in the production of those metals.

Under the microscope an intergranular fracture surface is bright, although the facets do not show the river markings characteristic of a cleavage fracture path. Section 6.7.1 (*d*) on p. 130 describes a simple experiment to demonstrate intergranular fracture in a solid.

6.3. *Ductile to brittle transition in fracture*

The fact that a crack nucleates in a crystal does not necessarily mean that it will propagate. Many solids can break in either a ductile or a brittle manner depending upon a number of variables—the temperature and the strain rate being two important factors. Sharp notches

Fig. 6.10.   Failure at sea of an American T/2 tanker.

in the material are also particularly effective in changing the fracture from ductile to brittle in certain materials.

This effect is of great practical importance, as in complex structures and under irregular service conditions a situation may arise which favours the propagation of a brittle crack, as for example in the case of many all-welded ships during the Second World War. Of 5000 U.S.A. merchant ships built, over 1000 developed cracks in the hull within three years of service, and some broke completely in two, since the continuity of metal across the welded joints allowed a crack to pass without interruption from one plate to the next (fig. 6.10). A riveted ship, through using more steel, due to the overlapping of the plates necessary in the construction, would not have been as susceptible to total failure, as a given crack could have been confined to a single plate of the hull.

One method of testing for brittleness in service is to take a small notched bar of the metal, supported at its ends, and to strike it behind the notch with a heavy pendulum. The energy absorbed in fracturing the test-piece is measured from the amplitude of the swing of the pendulum after it has fractured the specimen. The results obtained from these ' notch-impact ' tests fall into two broad classes, as illustrated in fig. 6.11, if the results obtained by testing at different

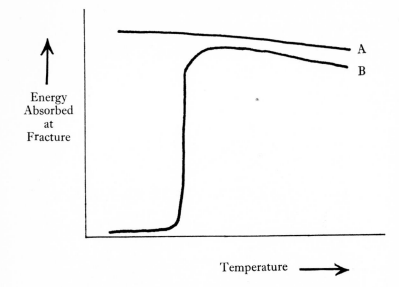

Fig. 6.11. Notch-impact fracture data for (A) f.c.c. metals and alloys, (B) materials showing a ductile–brittle transition temperature.

temperatures are plotted. Curve A, which shows a high energy absorption, and tough ductile behaviour at all temperatures, is typical of f.c.c. metals and alloys. Curve B, which shows tough behaviour at high temperatures, exhibits a transition over a narrow range of temperature to a brittle fracture, characterized by bright cleavage facets on the fracture surfaces of the specimen, and a low fracture energy recorded on the apparatus. Metals of b.c.c. structure fall into this category, and in particular structural steel can show this transition at ambient temperatures.

The transition in behaviour can be accounted for in the following way. As the temperature decreases, the yield stress of a metal

119

increases, since dislocation movement within the grains become more difficult. Metals possessing an f.c.c. structure show only a relatively small temperature dependence of the yield stress (fig. 6.12), but metals with a b.c.c. structure usually show a marked temperature dependence of yield (fig. 6.12). Many metals of h.c.p. crystal structure appear to behave in a similar way to f.c.c. materials, although there is some variation in their behaviour. In susceptible metals and

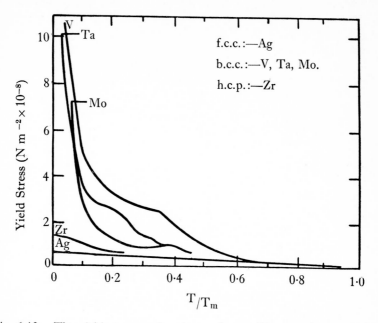

Fig. 6.12. The yield stresses of various polycrystalline metals as a function of temperature, expressed as the ratio of the test temperature to the melting temperature in °K.

alloys, if the temperature becomes low enough, the stress for plastic deformation may be so high that, before it is reached, the sample may fail in a brittle manner—i.e. the stress given by equation (6.4) may be reached below the stress for general plastic yield. The situation is illustrated in fig. 6.13 : it shows the variation of the yield stress and the brittle strength with temperature. In the case of b.c.c. metals and alloys these curves intersect at the ductile–brittle transition temperature. When it is deformed above this temperature the plastic yield stress is reached first, whereas below this temperature the brittle fracture stress is first exceeded. In f.c.c. materials the slope of the yield stress–temperature curve is not steep enough for it ever to intersect the brittle fracture stress curve (however low the temperature) so that alloys with this crystal structure never show this type of transition. Section 6.7.1 (*c*) on p. 129 describes a simple experiment to demonstrate the ductile to brittle transition in a solid.

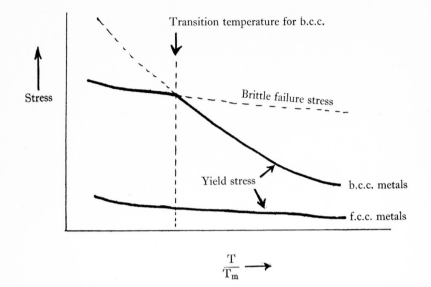

Fig. 6.13. Illustrating the transition in mechanical behaviour of b.c.c. metals with change of temperature. At low temperatures the brittle fracture strength is below the yield stress, whereas at high temperatures the yield stress is lower and plastic deformation occurs.

6.4. *High-temperature fracture*

When a metal is subjected to creep deformation (fig. 4.13), fracture is preceded by a period of progressively accelerating strain rate (tertiary creep). This effect results primarily from the development of intergranular cavities and cracks which ultimately lead to intergranular creep failure.

Creep fracture appears to initiate in two ways : (*a*) from wedge-type cracks (referred to as W-type cracks) which are formed characteristically at grain boundary triple points and (*b*) from small spherical cavities (referred to as r-type cavities) in the grain boundaries.

W-type cracks are observed particularly under conditions of creep at high stresses, and fig. 6.14 illustrates a typical mechanism which has been proposed to account for their formation. During creep, grain-boundary sliding takes place, and this results in stress-concentrations building up at triple points (i.e. at points where three grain boundaries meet in the structure). Grain boundaries lying approximately perpendicular to the direction of applied stress may tend to be wedged apart by grains such as A (fig. 6.14) tending to slide as the specimen elongates in creep. Such W-type cracks have been observed to open up during creep in a large number of pure metals and in a number of commercial alloys such as stainless steel and ' Nimonic 80 ' and ' Nimonic 90 ' which are special creep-resistant alloys developed for such applications as the blades of gas-turbine engines.

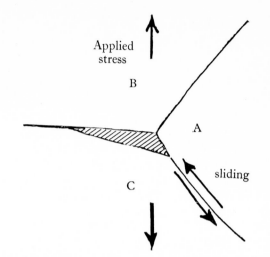

Fig. 6.14. Formation of a triple-point crack by grain boundary sliding; crystal A is being forced between grains B and C, thus wedging open the transverse grain boundary.

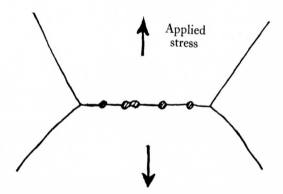

Fig. 6.15. Formation of grain–boundary voids in creep specimen.

r-type cavities, whose origins are still uncertain, tend to form under conditions of creep at low stresses and high temperatures, and in tensile creep tests they form predominantly in grain boundaries which make angles close to 90° to the tension axis (fig. 6.15).

122

During tertiary creep both W-type and r-type crack nuclei are observed to link up with their neighbours, so that the grain-boundary crack length progressively lengthens. Both types of cavity will grow at an accelerating rate, since the effective load-bearing cross-section of the specimen is reduced by their presence, so that the material is subjected to an ever-increasing stress. This accounts for the steep increase in tertiary creep rate prior to final rupture.

6.5. *Fatigue failure*

Metal fatigue has been estimated to account for about 90% of all metallurgical failures in service, so a great deal of research has taken place into the mechanism of this phenomenon, yet some of the details of it are still imperfectly understood. A metal may fail by fatigue if it is loaded to a stress which, if applied steadily, would be perfectly safe, yet if the load is varied either by pulsating it or fluctuating it, or by applying the load alternately in tension and then in compression (i.e. as an ' alternating load '), it may develop a crack and fail if the number of applications of the load is great enough—and this number may run into millions of cycles. This type of failure is encountered in many engine components such as gears, connecting rods and springs. Axles sometime fail by fatigue because any load which tends to *bend* the axle produces a stress which is a maximum in the skin of the axle. As the axle rotates, a given filament in the surface layer experiences stresses alternating from tensile to compressive in sign. Again, the spokes of a bicycle wheel experience a cycle of alternating tensile and compressive stress each time the wheel rotates under load : fatigue fracture of these components is not uncommon.

6.5.1. *Fatigue testing*

In the laboratory fatigue testing of materials, the stress is usually varied sinusoidally at a known maximum amplitude and is applied to a test-piece by bending, torsion, tension or compression at the rate of a few kilocycles per minute. A large number of specimens is prepared, and the number of stress cycles (N) endured before failure by each specimen at a given stress amplitude (S) is recorded and plotted on an $S-N$ diagram as shown in fig. 6.16. The curve generally has the form of A in fig. 6.16, with a ' knee ', or change of slope, in the curve at fatigue lives of the order 10^6 cycles. Some materials, notably many steels in the absence of a corrosive environment, give a curve of shape B (fig. 6.16), which indicates that the metal can withstand *indefinitely* the application of stress cycles below the value S_L, which is known as the ' fatigue limit '. However, such a limit cannot be defined for most non-ferrous metals and alloys, or for steels when they are used in corrosive conditions, as the $S-N$ curve never becomes horizontal (A), at least up to lives of the order 10^9 cycles.

6.5.2. *The nucleation and propagation of fatigue cracks*

Fatigue cracks are easily recognized from the appearance of the fracture : two distinct zones can usually be seen (fig. 6.17). A small crack forms, usually at a point of stress concentration in the component, and then slowly spreads across the material giving rise to the smooth region as on the right hand side of fig. 6.17. The unbroken

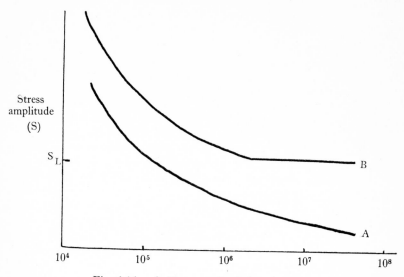

Fig. 6.16. *S–N* curves for fatigued metals.

part of the cross-section will eventually be unable to support the applied load and it will then fail suddenly by ductile or brittle fracture producing the second region of the fracture, which may be dull and fibrous or faceted.

During fatigue stressing, slip continues repeatedly to and fro, and where active slip bands meet the free surface thin tongues of metal (called ' extrusions ') are seen to appear (fig. 6.18) and narrow crevices (called ' intrusions ') spread down the slip plane from the surface. Thus small surface cracks lying on the slip plane form very early in the fatigue life (perhaps only 10% of the total life), and then for some reason they remain almost dormant until the final 10% of the fatigue life, when the crack turns out of the slip plane into a path roughly perpendicular to the tensile axis. The fatigue crack now advances a step at a time, once during each stress cycle, and the fracture surface becomes covered with ' ripple marking '—each ripple or striction marking successive positions of the crack front. The crack thus spreads at an increasing rate until the remaining cross-section is reduced sufficiently to be ruptured under a single tensile pull.

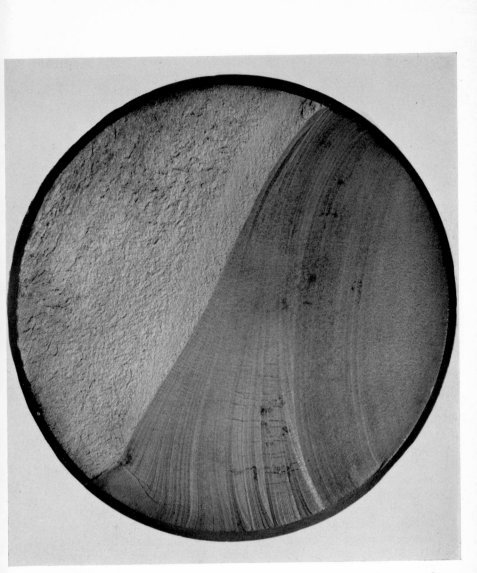

Fig. 6.17. A fatigue fracture surface upon a large steel shaft. (Courtesy of the British Engine Boiler and Electrical Insurance Co. Ltd.)

6.5.3. *Factors affecting fatigue behaviour*

Since the nucleation of fatigue failures occurs at the surface of the component, it follows that any alteration of surface properties must bring about a change in fatigue properties. The removal of machining marks and other irregularities on the surface of a component always leads to an improvement in fatigue life. It is found that the majority of service fatigue failures arise at points of surface stress concentration

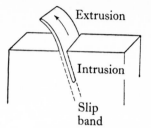

Fig. 6.18. Slip-band extrusion formation on the surface of a fatigued specimen.

such as keyways, fillets, lubrication holes or even at identification marks punched in the surface. These are the greatest hazard and may arise from bad workmanship or as a result of bad design.

Uniformly hardening the surface of the material to suppress the plastic processes improves the fatigue strength : this can be done in several ways. The surface can be *work-hardened* by emerying or by shot-peening (which involves allowing a stream of steel shot to impinge on the surface in a high-velocity jet) which also introduces compressive stresses into the surface layers of the metal. For steel components other common surface treatments which give rise to a hard layer in a state of compressive stress are nitriding and carburizing. Here nitrogen or carbon is allowed to diffuse into the surface of the alloy at elevated temperatures to produce a hard layer containing nitride or carbide phases in the material. The volume dilatation associated with the phase change is the source of internal compressive stress in the hardened layer.

The effect of alloying upon a metal is to increase the fatigue strength, and in stable alloys (e.g. solid-solution strengthened alloys) the increase in fatigue strength is directly proportional to the increase in static tensile strength. In unstable alloys, however, such as age-hardened material, fatigue stressing (as in creep) is found to enhance the rate of approach to equilibrium in local regions associated with the fatigue crack. The changes which occur appear to be associated with the enhanced diffusion brought about by the production of vacancies during the fatigue test.

This effect means that many alloys (for example the high-strength light alloys used for airframe construction), which have been developed to give a high resistance to static stresses by age-hardening heat treatments, have been found to exhibit disappointing fatigue strengths —the proportional increase in fatigue strength being much smaller than that in its static strength.

6.6. *Fracture of fibre-composite materials*

We have seen that an equation of the form of (6.3) will show at what applied stress a crack will spread in a solid. Whereas in a brittle solid such as glass γ equals the surface energy, for a material such as a ductile metal γ must be replaced by P (equation (6.4)) which is some 10^5 times greater than γ, because plastic flow at the tip of a crack is required to separate the atoms, which cannot be cleaved apart in a brittle way. The resultant stress σ is so large under these conditions that the material usually shows general plastic yield before the stress is high enough to extend the crack.

In strong alloys, dislocation motion is restricted and P (equation (6.4)) is reduced, so that as the strength of an alloy is increased there is increasing danger that it will become ' notch-sensitive ', that is,

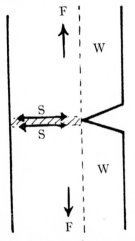

Fig. 6.19. Illustrating the development of transverse stresses (S) in a deformed notched bar.

liable to fail with low ductility by the propagation of any existing crack or notch. The yield strength is raised so much in these materials that the stored elastic energy becomes high enough to create fresh crack surfaces before general yield sets in.

In strong fibrous aggregates the free running of cracks is prevented by factors additional to plastic flow to the matrix material itself. Let us consider an aggregate consisting of a parallel array of strong fibres embedded in a soft plastic matrix subjected to a tensile stress parallel to the axis of the fibres. First consider a solid rod containing a transverse notch or crack (fig. 6.19) subjected to an axial tensile force F. The

127

volume of metal (shown shaded in the diagram) at the minimum cross-section will experience the greatest stress concentration and this region will start to extend plastically. In order to maintain its volume constant, however, this region will also have at the same time to contract transversely (i.e. in the plane of the crack). In an unnotched solid this transverse or lateral contraction which accompanies plastic extension is unimpeded, but in the present specimen the material forming the walls of the notch (marked W in fig. 6.19) will oppose this contraction, so that the shaded volume will experience transverse tensile forces parallel to the plane of the notch. Thus, ahead of any transverse crack or notch which is under axial tensile stress *transverse* tensile forces will develop, due to the specimen geometry.

If, in a fibre composite, the adhesion between the fibres and the matrix is low, the material is weak in a direction at right angles to the fibres, and if a crack starts to run in a direction perpendicular to the fibres, the lateral tensile stress will cause the fibre–matrix interface to pull apart, or ' decohere ', as shown in fig. 6.20. By this process the

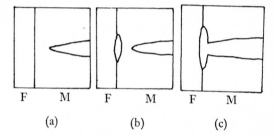

F   M   F   M   F   M

(a)   (b)   (c)

Fig. 6.20. Crack in ductile matrix (M) being deflected at fibre (F)–matrix interface.

crack is deflected along the weak interface, and it is thus rendered harmless as far as the desirable properties parallel to the fibres are concerned.

A second effect contributes to a high work to fracture in these materials, which makes fibre composites tough. Although the reinforcing fibres may fail when the composite is severely deformed, they will not do so all in one plane, and to separate the composite into halves it is necessary to pull the fibres out of the matrix as they break. Work must be done by the applied stress in pulling out the fibres against the holding force of the matrix, and this leads to a high absorption of energy in this type of material and thus high resistance to crack propagation under tensile forces.

Fibre-composite materials are thus ideally suited to carrying large tensile loads in one direction, but may not be as effective as other materials when more complex stress patterns are applied. Their application in practice therefore requires very careful engineering design in order to make full use of their potentialities, yet overcoming their deficiencies.

6.7. Simple experiments on fracture

6.7.1. Demonstrations of fracture processes

(a) Ductile fracture

Ordinary modelling-clay can be used to represent the behaviour of a very ductile metal. Tensile elongation and neck formation can be readily demonstrated.

Although it ruins the material for other purposes, it is possible to demonstrate the importance of inclusions in ductile fracture processes by preparing a series of rods of modelling-clay into which an increasing proportion of sand has been mixed as uniformly as possible. The elongation can be estimated by making two marks a few centimetres apart on the rods before deforming them, then, after fracture, fitting the two halves together and remeasuring the separation of the reference marks. The percentage elongation can then be calculated easily.

(b) Cleavage fracture

Pure zinc is a suitable metal to provide an example of cleavage fracture. If a rod of zinc is not available some metal can be melted in a crucible over a bunsen burner (in a well-ventilated room) and the molten metal poured into a short length of hard-glass tubing of at least 10 mm diameter. In this way a rod, perhaps 50 mm in length, can be prepared.

A sharp saw-cut, 2–3 mm deep, should now be made in the centre of the rod in a direction perpendicular to the length. If the lower half of the rod is now gripped in a vice and the upper half struck above the notch with a hammer, the specimen can be cleaved into halves. Any tendency for ductile behaviour can be suppressed by putting the zinc rod into the freezing compartment of a refrigerator for a short while before fracturing it.

Examine the bright facets on the fracture surface under a magnifying glass or low-power microscope. Although these are not intergranular facets, the grain size can be readily distinguished. Why is this?

(c) Ductile-to-brittle transition

If the above experiment is repeated on zinc which has been in boiling water and is close to 100°C when deformed (this involves considerable practice in getting the rod from the water-bath into the vice with minimum delay), completely ductile behaviour will be encountered.

A simpler demonstration of such a transition in fracture under variation in strain rate rather than in temperature can be made using silicone ' bouncing putty '. If a short rod of this is pulled very slowly it will behave like ordinary moulding clay and exhibit a ductile mode of fracture (with 100% reduction in area).

If a similar rod is sharply extended it will fracture on a plane approximately perpendicular to the maximum tensile stress, showing very little overall tensile elongation. This experiment, although it provides a striking mechanical analogy to the behaviour of many metals and alloys, cannot be equated in its behaviour on an atomic scale.

(d) Intergranular fracture

The grain size of most metals is too fine to resolve without the aid of a microscope. A simple analogy of intergranular fracture is readily made using a sample of expanded polystyrene (such as is used for ceiling tiles, in packing and for thermal insulation).

If a bar of this material is fractured, its surface will be seen to be faceted as the fracture takes place around the individual bubbles of the foam produced when this material is manufactured. We have already seen the close analogy between a foam and an array of metal grains, and this fracture surface also is exactly analogous to that obtained when alloys fracture intergranularly. This is found in practice if the alloy contains a low melting-point phase at the grain boundaries and it is deformed at elevated temperatures. The low-ductility intergranular failure is termed ' hot-shortness '.

(e) Fracture of a fibre-composite

Many natural materials are essentially fibre-strengthened composite materials, and a simple experiment of how such a composite can prevent a crack running freely can be carried out using a length of bamboo wood.

Make a sharp transverse notch in the rod and then bend it as far as possible, noting closely what happens.

Splinters may form near to the notch, but no cracks will run into the material and it will not snap into two pieces. In this material, strong bundles of cellulose fibres are separated by lignin-based material at relatively weak, yielding interfaces.

6.7.2. Work of fracture

The work of fracture (P, equation (6.4)) can be estimated crudely for a ductile material such as aluminium in the following way.

Test specimens in the form of ribbons 10–20 mm wide and 100 mm in length should be cut from aluminium foil of known thickness. Two clamps are required to hold the ends of the ribbon : these can be small grips which can be tightened by means of a wing-nut, or each can consist of a pair of brass blocks drilled and tapped so that they can be clamped together by a pair of bolts.

One clamp is held in a vice, and the clamp at the other end attached to a spring balance of approximate capacity 2 kg. The specimen is pulled by hand to fracture, noting the load at which this occurs. (The width of the specimen is adjusted until the fracture force is just within the capacity of the spring balance.)

The experiment consists of measuring the fracture load of the ribbons in the edges of which sharp slits have been cut with a razor blade, perpendicular to the axis, midway between the ends. One should observe the load at which fracture begins at the root of the notch, the extent to which it spreads before rapid fracture sets in, and the load at rapid fracture. From equation (6.4) it is seen that a graph of σ^2 against the reciprocal of the crack length should have a slope of $E.P$, where $E =$ Young's modulus (7×10^{10} N m^{-2} for aluminium).

Supplementary reading list

The Bakerian Lecture, 1963 : ' Fracture ' by A. H. Cottrell. (*Proc. Roy. Soc.* A, **276** (1963), 1.)

APPENDIX

LISTS of apparatus and materials required for the various exercises and experiments described in the text.

Chapter 1
Polystyrene spheres
Polystyrene cement

Chapter 2
Cyclohexanol
Ice
Stearic acid
Microscope slides and cover slips
Glass tube 3 cm diameter, 15 cm length.   Rubber stopper to fit
Tall glass beaker
Bunsen burner
Electric soldering iron
Hacksaw
Metal ruler

Chapter 4
0·02–0·03 kg of lead
Liquid detergent
0·05 kg of tin
Lead wire, 1 mm diameter
Low-power microscope
Crucible
Bunsen burner
Smooth steel plate
Medium-size plastic developing dish
Fine glass nozzle
Spatula
Small glass rake
Compressed air supply
Microscope slides
Heat-resistant glass tube, 5 mm diameter
Vice
Soft-glass tube, 20 mm diameter, 200 mm length
Water pump

Metre rule
Balance weights
Tin can, rods
Clock, thermometer, beaker, clamps

Chapter 5

Copper rod, 5 mm diameter, 100 mm length
Bunsen burner
Tongs
Hammer
Anvil or vice
Beaker of water
File
' Silver steel ' rods, 3–5 mm diameter, 150 mm long
Emery paper (00 or 000 grade)

Chapter 6

Modelling-clay
Sand (few grammes)
0·05 kg of zinc
Silicone ' putty '
Expanded polystyrene block or rod (e.g. ceiling tiles)
Bamboo rod
Aluminium foil
Hard-glass tube, 10 mm diameter, 500 mm long
Hacksaw
Vice
Hammer
Refrigerator
Low-power microscope or hand magnifier
Beaker of boiling water
Spring balance (2 kg capacity)
Razor blade
Clamps

INDEX

134

Metallurgical microscope   25–27
Metallurgy (definition)   vii–ix
Multiple slip   71

Nickel
  TD nickel   104
  nickel–copper alloys   41, 43
Normalizing   59
Notch-impact testing   119
Nucleation   20–21

Over-ageing   100

Pearlite   58–59
Peritectic   51–52, 57–58
Peritectoid   55
Phase diagrams   42
  copper–nickel   43
  bismuth–cadmium   48
  iron–carbon   58
  lead–tin   49
Plastic deformation   68–78, 88 *et seq.*
Point-defect hardening   88–91
Polymorphism   16
Powder metallurgy   103
Precipitation hardening   98–101
Proof stress (definition)   70

Radiation hardening   89–91
Recrystallization   79, 83
Relative valency factor   42
Richardson triangle   vii–ix
River markings   117

S.A.P.   103–105
Scanning electron microscope   34–35
Shear modulus (definition)   67

Slip   70–71, 74–77, 82
Solder   49
Solidification   20–23, 37–39
Solid solutions   40–44, 96–98
— state phase transformations   53–62
Solute hardening   96–97
Specific modulus (definition)   68
Steels
—, hardening (experiment)   106–107
—, microstructures   57–62
Strain (definition)   65
Stress (definition)   64

Tempering of steel   61–62, 107
Tin   74, 83–84
  tin–lead alloys   49
Tolman effect   17
Twins, annealing   42
—, deformation   72–74, 83–84

Ultimate tensile stress   110
Unit cell   11
Uranium   73
Van der Waals bonding   7

Widmannstatten structure   2, 54–55
Work-hardening   69, 77–78, 91–94

X-ray diffraction   8–11

Yield point   97
— stress (definition)   69
Young's modulus (definition)   67

Zinc   23, 129

THE WYKEHAM SCIENCE SERIES

for schools and universities

Price per book for the Science Series **20s.—£1.00 net** *in U.K. only*

THE WYKEHAM TECHNOLOGICAL SERIES

for universities and institutes of technology

Price per book for the Technological Series **25s.—£1.25 net** *in U.K. only*